Python and R for the
Modern Data Scientist
The Best of Both Worlds

Rick J. Scavetta and Boyan Angelov

Beijing · Boston · Farnham · Sebastopol · Tokyo

Python and R for the Modern Data Scientist

by Rick J. Scavetta and Boyan Angelov

Copyright © 2021 Boyan Angelov and Rick J. Scavetta. All rights reserved.

Published by O'Reilly Media, Inc., 1005 Gravenstein Highway North, Sebastopol, CA 95472.

O'Reilly books may be purchased for educational, business, or sales promotional use. Online editions are also available for most titles (*http://oreilly.com*). For more information, contact our corporate/institutional sales department: 800-998-9938 or *corporate@oreilly.com*.

Acquisitions Editor: Michelle Smith
Development Editor: Angela Rufino
Production Editor: Katherine Tozer
Copyeditor: Tom Sullivan
Proofreader: Piper Editorial Consulting, LLC

Indexer: Sam Arnold-Boyd
Interior Designer: David Futato
Cover Designer: Karen Montgomery
Illustrator: Kate Dullea

July 2021:   First Edition

Revision History for the First Edition
2021-06-22:   First Release

See *http://oreilly.com/catalog/errata.csp?isbn=9781492093404* for release details.

The O'Reilly logo is a registered trademark of O'Reilly Media, Inc. *Python and R for the Modern Data Scientist*, the cover image, and related trade dress are trademarks of O'Reilly Media, Inc.

The views expressed in this work are those of the authors, and do not represent the publisher's views. While the publisher and the authors have used good faith efforts to ensure that the information and instructions contained in this work are accurate, the publisher and the authors disclaim all responsibility for errors or omissions, including without limitation responsibility for damages resulting from the use of or reliance on this work. Use of the information and instructions contained in this work is at your own risk. If any code samples or other technology this work contains or describes is subject to open source licenses or the intellectual property rights of others, it is your responsibility to ensure that your use thereof complies with such licenses and/or rights.

978-1-492-09340-4

[LSI]

To my parents for giving me the best possible start in life. To my wife for being my rock. To my children for having the brightest dreams of the future.

–Boyan Angelov

For all of us ready, willing, and able to perceive the world through a wider lens, eliminating "the other" in our midst.

–Rick Scavetta

Table of Contents

Part III.   Bilingualism II: The Modern Context

Part IV.   Bilingualism III: Becoming Synergistic

Preface

Why We Wrote This Book

We want to show data scientists why being more aware, informed, and deliberate about their tools is an optimal strategy for increased productivity. With this goal in mind, we didn't write a bilingual dictionary (well, not *only*—you'll find that handy resource in the Appendix). Ongoing discussions about Python versus R (the so-called "language wars") have long since ceased to be productive. It recalls, for us, Maslow's hammer: "if all you have is a hammer, everything looks like a nail." It's a fantasy worldview set in absolutes, where one tool offers an all-encompassing solution. Real-world situations are context-dependent, and a craftsperson knows that tools should be chosen appropriately. We aim to showcase a new way of working by taking advantage of all the great data science tools available, regardless of the language they are written in. Thus we aim to develop both how the modern data scientist thinks and works.

We chose the word *modern* in the title not just to signify novelty in our approach. It allows us to take a more nuanced stance in how we discuss our tools. What do we mean by *modern data science*? Modern data science is:

Collective
> It does not exist in isolation. It's integrated into wider networks, such as a team or organization. We avoid jargon when it creates barriers and embrace it when it builds bridges (see "Technical Interactions" on page x).

Simple
> We aim to reduce unnecessary complexity in our methods, code, and communications.

Accessible
> It's an open design process that can be evaluated, understood, and optimized.

Generalizable
  Its fundamental tools and concepts are applicable to many domains.

Outward looking
  It incorporates, is informed by, and is influenced by developments in other fields.

Ethical and honest
  It's people-oriented. It takes best practices for ethical work, as well as a broader view of its consequences, for communities and society, into account. We avoid hype, fads, and trends that only serve short-term gains.

However the actual job description of a data scientist evolves in the coming years, we can expect that these timeless principles will provide a strong foundation.

Technical Interactions

Accepting that the world is more extensive, more diverse, and more complex than any single tool can serve presents a challenge that is best addressed directly and early.

This broadened perspective results in an increase in *technical interactions*. We must consider the programming language, packages, naming conventions, project file architecture, integrated development environments (IDEs), text editors, and on and on that will best suit the situation. Diversity gives rise to complexity and confusion.

The more diverse our ecosystem becomes, the more important it is to consider whether our choices act as bridges or barriers. We must always strive to make choices that *build bridges* with our colleagues and communities and avoid those that *build barriers* that isolate us and make us inflexible. There is plenty of room to contain all the diversity of choices we'll encounter. The challenge in each situation is to make choices that balance personal preference and communal accessibility.

This challenge is found in all technical interactions. Aside from tool choice (a "hard" skill), it also includes communication (a "soft" skill). The content, style, and medium of communication, to name just a few considerations, also act as bridges or barriers to a specific audience.

Becoming bilingual in both Python and R is a step toward building bridges among members of the wider data science community.

Who This Book Is For

This book aims at data scientists at the intermediate stage of their careers. As such, it doesn't attempt to *teach* data science. Nonetheless, early-career data scientists will also benefit from this book by learning what's possible in a modern data science context before committing to any topic, tool, or language.

Our goal is to bridge the gap between the Python and R communities. We want to move away from a tribal, "us versus them" mentality and toward a unified, productive community. Thus, this book is for those data scientists who see the benefit of expanding their skill set and thereby their perspectives and the value that their work can add to all variety of data science projects.

It's negligent to ignore the powerful tools available to us. We strive to be open to new, productive ways of achieving our programming goals and encourage our colleagues to get out of their comfort zone.

In addition, Part II and the Appendix also serve as useful references for those moments when you just need to quickly map something familiar in one language onto the other.

Prerequisites

To obtain the best value from this book, we assume the reader is familiar with at least one of the main programming languages in data science, Python and R. A reader with knowledge of a closely related one, such as Julia or Ruby, can also derive good value.

Basic familiarity with general areas of data science work, such as data munging, data visualization, and machine learning is beneficial, but not necessary, to appreciate the examples, workflow scenarios, and case study.

How This Book Is Organized

We've organized this book as if we're learning a second spoken language as an adult.

In Part I we begin by going back in time to the origins of the two languages and show how this has influenced the current state by covering key breakthroughs. In our analogy with spoken languages, this helps provide a bit of context as to why we have quirks such as irregular verbs and plural endings. Etymology is interesting and helps you gain an appreciation of a language, like the seemingly endless forms of plural nouns in German, but it's certainly not essential for speaking.[1] If you want to get right into the languages, skip straight to Part II.

Part II provides a deeper dive into the dialects of both languages by offering a mirrored perspective. First we will cover how a Python user should approach work with R, and then the other way around. This will expand not only your skill set but also your way of thinking as you appreciate how each language operates.

---

1 Etymology is the study of word origins and meanings.

In this part, we'll treat each language separately as we start to become bilingual. Just like becoming bilingual in a spoken language, we need to resist two defeating urges. The first urge is to point out how much more straightforward, or more elegant, or in some way "better," something is in our mother tongue. Congratulations to you, but that's not the point of learning a new language, is it? We're going to learn each language in its own right. Although we'll point out comparisons as we go along, they'll help us deal with our native-language baggage.

The second urge is to constantly try to interpret *literally* and *word for word* between two languages. This prevents us from *thinking* (or even *dreaming*) in the new language, and sometimes it's just not possible! Examples I like to use are phrasing such as *das schmeckt mir* in German, or *ho fame* in Italian, which translate literally very poorly as "that tastes to me" (That tastes good) and "I have hunger" (I'm hungry). The point is, different languages allow for different constructs. This gives us new tools to work with and new ways to think, once we realize that we can't map everything 1:1 onto our previous knowledge. Think of these chapters as our first step to mapping your knowledge of one language onto the other.

Part III covers the modern context of language applications. This includes a review of the broad ecosystem of open source packages as well as the variety of workflow-specific methods. This part will demonstrate when one language is preferred and why, although they'll still be separate languages at this point. This will help you to decide which language to use for parts of a large data science project.

In spoken languages, *lost in translation* is a real thing. Some things just work better in one language. In German, *mir ist heiß* and *ich bin heiß* are both "I'm hot" in English, but a German speaker will distinguish hotness from the weather versus physique. Other words like *Schadenfreude*, a compound word from "schaden" (damage) and "freude" (pleasure) meaning to take pleasure in someone's difficulties, or *Kummerspeck*, a compound word from "kummer" (grief) and "speck" (bacon) referring to the weight gained due to emotional eating, are just so perfect there's no use in translating them.

Part IV details the modern interfaces that exist between the languages. First, we became bilingual, using each language in isolation. Then, we identified how to choose one language over another. Now, we'll explore tools that take us from separate and interconnected Python and R scripts to single scripts that weave the two languages together in a single workflow.

The real fun starts when you're not just bilingual, but working within a bilingual community. Not only can you communicate in each language independently, but you can also combine them in novel ways that only other bilingual speakers will appreciate and understand. Bilingualism doesn't just provide access to a new community but also creates in itself a new community. For purists, this is pure torture, but I hope we've moved beyond that. Bilinguals can appreciate the warning "The *Ordnungsamt* is monitoring *Bergmannkiez* today." Ideally you're not substituting words because you've forgotten them, but because it's the best choice for the situation. There's no great translation of Orgnungsamt (regulatory agency?) and Bergmannkiez is a neighborhood in Berlin that shouldn't be translated anyways. Sometimes words in one language more easily convey a message, like *Mundschutzpflicht*, the obligatory wearing of face masks during the coronavirus pandemic.

Finally, Chapter 7 consists of a case study that will outline how a modern data science project can be implemented based on the material covered in this book. Here, we'll see all the previous sections come together in one workflow.

Let's Talk

The field of data science is continuously evolving, and we hope that this book will help you navigate easily between Python and R. We're excited to hear what you think, so let us know how your work has changed! You can contact us via the companion website for the book (*https://www.moderndata.design*). There you'll find updated extra content and a handy Python/R bilingual cheat sheet.

Conventions Used in This Book

The following typographical conventions are used in this book:

Italic
> Indicates new terms, URLs, email addresses, filenames, and file extensions.

`Constant width`
> Used for program listings, as well as within paragraphs to refer to program elements such as variable or function names, databases, data types, environment variables, statements, and keywords.

`Constant width bold`
> Shows commands or other text that should be typed literally by the user.

`Constant width italic`
> Shows text that should be replaced with user-supplied values or by values determined by context.

 This element signifies a tip or suggestion.

This element signifies a general note.

 This element indicates a warning or caution.

Using Code Examples

Supplemental material (code examples, exercises, etc.) is available for download at *https://github.com/moderndatadesign/PyR4MDS*.

If you have a technical question or a problem using the code examples, please send email to *bookquestions@oreilly.com*.

This book is here to help you get your job done. In general, if example code is offered with this book, you may use it in your programs and documentation. You do not need to contact us for permission unless you're reproducing a significant portion of the code. For example, writing a program that uses several chunks of code from this book does not require permission. Selling or distributing examples from O'Reilly books does require permission. Answering a question by citing this book and quoting example code does not require permission. Incorporating a significant amount of example code from this book into your product's documentation does require permission.

We appreciate, but generally do not require, attribution. An attribution usually includes the title, author, publisher, and ISBN. For example: "*Python and R for the Modern Data Scientist* by Rick J. Scavetta and Boyan Angelov (O'Reilly). Copyright 2021 Boyan Angelov and Rick J. Scavetta, 978-1-492-09340-4."

If you feel your use of code examples falls outside fair use or the permission given above, feel free to contact us at *permissions@oreilly.com*.

O'Reilly Online Learning

 For more than 40 years, *O'Reilly Media* has provided technology and business training, knowledge, and insight to help companies succeed.

Our unique network of experts and innovators share their knowledge and expertise through books, articles, and our online learning platform. O'Reilly's online learning platform gives you on-demand access to live training courses, in-depth learning paths, interactive coding environments, and a vast collection of text and video from O'Reilly and 200+ other publishers. For more information, visit *http://oreilly.com*.

How to Contact Us

Please address comments and questions concerning this book to the publisher:

O'Reilly Media, Inc.
1005 Gravenstein Highway North
Sebastopol, CA 95472
800-998-9938 (in the United States or Canada)
707-829-0515 (international or local)
707-829-0104 (fax)

We have a web page for this book, where we list errata, examples, and any additional information. You can access this page at *https://oreil.ly/python-and-r-data-science*.

Email *bookquestions@oreilly.com* to comment or ask technical questions about this book.

For news and information about our books and courses, visit *http://oreilly.com*.

Find us on Facebook: *http://facebook.com/oreilly*

Follow us on Twitter: *http://twitter.com/oreillymedia*

Watch us on YouTube: *http://www.youtube.com/oreillymedia*

Acknowledgments

The authors acknowledge the contribution of many individuals who have helped make this book possible.

At O'Reilly, we thank Michelle Smith, a senior content acquisitions editor with unparalleled passion, breadth of knowledge and foresight with whom we were fortunate enough to work with. We thank Angela Rufino, our content development editor for keeping us on track during the writing process and lifting up our spirits with a literal wall of action heroes and kind words of encouragement. We are grateful to Katie Tozer, our production editor, for her patience and fastidious treatment of our manuscript. We are grateful to Robert Romano and the design team at O'Reilly. They not only aided in redrawing figures but also selected, as per our wishes, a vibrant, commanding and truly impressive colossal squid for the cover! We also thank Chris Stone and the engineering team for technical help.

A special thanks goes out to the countless unseen individuals working behind the scenes at O'Reilly. We appreciate the amount of effort needed to make excellent and relevant content available.

We are also indebted to our technical reviewers, who gave generously of both their time and insightful comments borne from experience: Eric Pite and Ian Flores at RStudio, our fellow O'Reilly authors Noah Gift and George Mount, and the impeccable author Walter R. Paczkowski. Your comments were well received and improved the book immensely.

Rick also thanks all his students, both online and in-person, from the past 10 years. Every chance to pass on knowledge and understanding reaffirmed the value of teaching and allowed, however slight, a contribution to the great scientific endeavor. Rick is also thankful for the dedicated administrative support that allows him to maintain an active relationship with primary scientists around the world.

Finally, we extend a heartfelt thanks to not only Python and R developers but also to the broad, interconnected community of open source developers. Their creativity, dedication, and passion are astounding. It is difficult to consider how the data science landscape would look without the collective efforts of thousands of developers working together, crossing all borders, and spanning decades. Almost nothing in this book would be possible without their contributions!

## PART I

Discovery of a New Language

To get things started, we'll review the history of both Python and R. By comparing and contrasting these origin stories, you'll better appreciate the current state of each language in the data science landscape. If you want to get started with coding, feel free skip ahead to Part II.

## CHAPTER 1

In the Beginning

Rick J. Scavetta

We would like to begin with a great first sentence, like "It was the best of times, it was the worst of times…" but honestly, it's just the best of times—data science is flourishing! As it continues to mature, it has begun to splinter into niche topics, as many disciplines do over time. This maturity is the result of a long journey that began in the early days of scientific computing. It's our belief that knowing some of Python's and R's origin stories will help you to appreciate how they differ in today's environment and, thus, how to get the most out of them.

We're not going to pretend to be science historians, that niche group of academics who trace the circumstances of great discoveries and personalities. What we can do is offer a highlight reel of where Python and R come from and how that led us to our current situation.

The Origins of R

Whenever I think about R, I'm reminded of FUBU, a streetwear company founded back in the 1990s. The name is an acronym that I immediately fell in love with: *For Us, By Us*. FUBU meant community; it meant understanding the needs and desires of your people and making sure you served them well. *R is FUBU.*[1] By the end of this chapter, I'm sure you'll feel the same way. Once we acknowledge that R is FUBU, it starts to make a lot more sense.

We can trace the origins of R right back to the now legendary Bell Laboratories in New Jersey. In 1976, development of the statistical programming language S was being spearheaded by John Chambers. A year later, Chambers published

---

1 Well, OK, more like *For Statisticians, By Statisticians*, but FSBS doesn't have the same ring to it.

Computational Methods for Data Analysis (John Wiley & Sons) and his colleague John Tukey, also at Bell Laboratories, published *Exploratory Data Analysis* (Addison-Wesley). In 1983, Chambers et al. published *Graphical Methods for Data Analysis* (CRC Press). These books provided the framework to develop a computational system that would allow a statistician to not only explore, understand, and analyze their data, but also to communicate their results. We're talking about an all-star FUBU lineup here! Coauthors of Chambers included both Tukey's cousin Paul A. Tukey and William Cleveland. Cleveland's empirical experiments on perception, summarized in two insightful books, continue to inform the broader field of data visualization to this day. Among their many contributions to scientific computing and statistics, Tukey developed novel visualizations, like the oft misunderstood box and whiskers plot, and Cleveland developed the LOESS (Locally Weighted Scatterplot Smoothing) method for nonparametric smoothing.

We begin with S because it laid the foundations for what would eventually become R. The nuggets of information in the previous paragraph tell us quite a bit about S's—and R's—foundations. First, statisticians are very literal people (S, get it?). This is a pretty helpful trait. Second, statisticians wanted a FUBU programming language specializing in data analysis. They weren't interested in making a generalist programming language or an operating system. Third, these early books on computational statistics and visualization are, simply put, stunning examples of pedagogical beauty and precise exposition.[2] They have aged surprisingly well, despite the obviously dated technology. I'd argue that these books planted the seed for how statisticians, and the R community in particular, approached technical communication in an open, clear, and inclusive manner. This, I believe, is an outstanding and distinctive hallmark of the R community that has deep roots. Fourth, the early emphasis on graphical methods tells us that S was already concerned with flexible and efficient data visualizations, necessary for both understanding data and communicating results. So S was about getting the most important things done as easily as possible, and in a true FUBU way.

The original distribution of S ran on Unix and was available for free. Eventually, S became licensed under an implementation titled S-PLUS. This prompted another open source and free implementation of S by Ross Ihaka and Robert Gentleman at the University of Auckland in 1991. They called this implementation R, for the initials of their first names, as a play on the name S, and in keeping with the tradition of naming programming languages using a single letter. The first official stable beta release of R v1.0.0 was available on February 29, 2000. In the intervening years, two important developments occurred. First, CRAN (*https://oreil.ly/HIpY7*), the Comprehensive R Archive Network, was established to host and archive R packages on mirrored servers. Second, the R Core Team was also established. This group of volunteers

---

2 With the possible exception of *Computational Methods for Data Analysis*, which I admit to not having read.

(*https://oreil.ly/Zjrvw*) (which currently consists of 20 members) implements base R, including documentation, builds, tests, and releases, plus the infrastructure that makes it all possible. Notably, some of the original members are still involved, including John Chambers, Ross Ihaka, and Robert Gentleman.

A lot has happened since R v1.0.0 in 2000, but the story so far should already give you an idea of R's unique background as a FUBU statistical computing tool. Before we continue with R's story, let's take a look at Python.

The Origins of Python

In 1991, as Ross Ihaka and Robert Gentleman began working on what would become R, Guido van Rossum, a Dutch programmer, released Python. Python's core vision is really that of one person who set out to address common computing problems at the time. Indeed, van Rossum was lovingly referred to as the benevolent dictator for life (BDFL) for years, a title he gave up when he stepped down from Python's Steering Council in 2018.

We saw how S arose out of the need for statisticians to perform data analysis and how R arose from the need for an open source implementation, so what problem was addressed by Python? Well, it wasn't data analysis—that came much later. When Python came on the scene, C and C++, two low-level programming languages, were popular. Python slowly emerged as an interpreted, high-level alternative, in particular after Python v2 was released in 2000 (the same year R v1.0.0 was released). Python was written with the explicit purpose to be, first and foremost, an easy to use and learn, widely adopted programming language with simple syntax. And it has succeeded in this role very well!

This is why you'll notice that, in contrast to R, Python is everywhere and is incredibly versatile. You'll see it in web development, gaming, system administration, desktop applications, data science, and so on. To be sure, R is capable of much more than data analysis, but remember, R is FUBU. If R is FUBU, Python is a Swiss Army knife. It's everywhere and everyone has one, but even though it has many tools, most people just use a single tool on a regular basis. Although data scientists using Python work in a large and varied landscape, they tend to find their niche and specialize in the packages and workflows required for their work instead of exploiting all facets of this generalist language.

Python's widespread popularity within data science is not entirely due to its data science capabilities. I would posit that Python entered data science by partly riding on the back of existing uses as a general-purpose language. After all, getting your foot in the door is halfway inside. Analysts and data scientists would have had an easier time sharing and implementing scripts with colleagues involved in system administration and web development because they already knew how to work with Python scripts.

This played an important role in Python's widespread adoption. Python was well suited to take advantage of high-performance computing and efficiently implement deep learning algorithms. R was, and perhaps still is, a niche and somewhat foreign language that the wider computing world didn't really get.

Although Python v2 was released in 2000, a widely adopted package for handling array data didn't take root until 2005, with the release of NumPy. At this time, SciPy, a package that has, since 2001, provided fundamental algorithms for data science (think optimization, integration, differential equations, etc.), began relying on NumPy data structures. SciPy also provides specialized data structures such as k-dimensional trees.

Once the issue of a standard package for core data structures and algorithms was settled, Python began its ascent into widespread use in scientific computing. The low-level NumPy and SciPy packages laid the foundation for high-level packages like pandas in 2009, providing tools for data manipulation and data structures like data frames. This is sometimes termed the *PyData stack*, and it's when the ball really got rolling.

The Language War Begins

The early 2000s set the stage for what some would later refer to as the *language wars*. As the PyData stack started to take shape, milestones in both Python and R began to heat things up. Four stand out in particular.

First, in 2002, BioConductor (*https://www.bioconductor.org*) was established as a new R package repository and framework for handling the burgeoning (read absolute explosion of) biological data in its myriad forms. Until this point, bioinformaticians relied on tools like MATLAB and Perl (along with classic command-line tools and some manual web-interface tools). MATLAB is still favored in specific disciplines, like neuroscience. However, Perl has been mostly superseded by BioConductor. BioConductor's impact on bioinformatics is hard to overstate. Not only did it provide a repository of packages for dealing with remote genetic sequence databases, expression data, microarrays, and so on, it also provided new data structures to handle genetic sequences. BioConductor continues to expand and is deeply embedded within the bioinformatics community.

Second, in 2006 the IPython package was released. This was a groundbreaking way to work on Python in an interactive notebook environment. Following various grants beginning in 2012, IPython eventually matured into the Jupyter Project (*https://jupyter.org*) in 2014, which now encompasses the JupyterLab IDE. Users often forget that Jupyter is short for "Julia, Python, and R" because it's very Python-centric. Notebooks have become a dominant way of doing data science in Python, and in 2018

Google released Google Colab (*https://oreil.ly/O1krw*), a free online notebook tool. We'll dig into this in Chapter 3.

Third, in 2007, Hadley Wickham published his PhD thesis, which consisted of two R packages that would fundamentally change the R landscape. The first, reshape, laid the foundations for what would later become formalized as the Tidyverse (*https://www.tidyverse.org*) (more on this later). Although reshape has long since been retired, it was the first glimpse into understanding how data structure influences how we think about and work with our data.[3] The second, ggplot2, is an implementation of the seminal book by Leland Wilkinson et al., *The Grammar of Graphics* (Springer), and provided intuitive, high-level plotting that greatly simplified previously existing tools in R (more on this in Chapter 5).

Finally, Python v3 was released in 2008. For years the question persisted as to which version of Python to use, v2 or v3. That's because Python v3 is backward-incompatible.[4] Luckily, this has been resolved for you since Python v2 was retired in 2020. Surprisingly, you can still buy a new MacBook Pro after that date with Python 2 preinstalled because legacy scripts still rely on it. So Python 2 lives on still.

The Battle for Data Science Dominance

By this point both Python and R had capable tools for a wide variety of data science applications. As the so-called "language wars" continued, other key developments saw each language find its niche.

Both Python and R were wrapped up in specific *builds*. For Python this was the Anaconda distribution, which is still in popular use (see Chapter 3). For R, Revolution Analytics, a data science software developer, produced Revolution R Open. Although their R build was never widely adopted by the community, the company was acquired by Microsoft, signaling strong corporate support of the R language.

In 2011, the Python community foresaw the boom in machine learning with the release of the scikit-learn package. In 2016, this was followed by the release of both TensorFlow and Keras for deep learning, also with a healthy dose of corporate support. This also highlights Python's strength as a high-level interpreter sitting on top of high-performance platforms. For example, you'll find Amazon Web Services (AWS) Lambda for massive highly concurrent programming, Numba for high-performance computing, and the aforementioned TensorFlow (*https://www.tensorflow.org*) for highly optimized C++. With its widespread adoption outside of data science, it's no

---

3 I would argue that we can trace this relationship back to the early days of R, as evidenced by formula notation and various built-in datasets. Nonetheless, a consistent and intuitive framework was lacking.

4 This has even moved some prominent developers to voice an aversion to the eventual development of Python 4.0. How Python develops will be exciting to watch!

surprise that Python gained a reputation for deploying models in a way that R could not.

2011 also saw the release of RStudio IDE (*https://rstudio.com*) by the eponymous company, and over the next few years the R community began to converge on this tool. At this point, to use R is, in many regards, to use RStudio. The influence RStudio has on promoting R as a programming language suitable for a wide variety of data-centric uses is also important to note.

While all of this was happening, a growing segment of the R community began to move toward a suite of packages, many of which were authored or spearheaded by Hadley Wickham, that began to reframe and simplify typical data workflows. Much of what these packages did was to standardize R function syntax, as well as input and output data storage structures. Eventually the suite of packages began to be referred to colloquially as the "Hadleyverse." In a keynote speech at the useR! 2016 conference at Stanford University, Wickham did away with this, igniting digital flames to burn up his name and coining the term "Tidyverse." Since Wickham joined RStudio, the company has been actively developing and promoting the Tidyverse ecosystem, which has arguably become the dominant dialect in R. We'll explore this in more depth in Chapter 2.

We can imagine that R contains at least two "paradigms," or "dialects." They can be mixed, but each has its own distinct flavor. Base R is what most R has been and, probably, still is. Tidyverse reimagines base R in a broad, all-encompassing universe of packages and functions that play well together, often relying on *piping*,[5] and has a preference for data frames.[6] I would argue that BioConductor provides yet another dialect, which is focused on a specific discipline, bioinformatics. You'll no doubt find that some large packages may contain enough idiosyncrasies that you may consider them a dialect in their own right, but let's not go down that rabbit hole. R is now at the threshold where some users know (or are taught) only the Tidyverse way of doing things. The distinction between base and Tidyverse R may seem trivial, but I have seen many new R learners struggle to make sense of why the Tidyverse exists. This is partly because years of base R code is still in active use and can't be ignored. Although Tidyverse advocates argue that these packages make life much easier for the beginner, competing dialects can cause unnecessary confusion.

We can also imagine that Python contains distinct dialects. The *vanilla* installation of Python is the bare-bones installation, and operates differently from an environment

---

5 That is, using an output of one function as the input of another function.

6 Python users might not be familiar with the term *base*. This means only the built-in functionality of the language without any additional package installations. Base R itself is well equipped for data analysis. In Python, a data scientist would import the PyData stack by default.

that has imported the PyData stack. For the most part, data scientists operate within the PyData stack, so there's less confusion between dialects.

A Convergence on Cooperation and Community-Building

For a time, it seemed that the prevailing attitude in the language wars was an us versus them mentality. A look of disdain glancing at a person's computer screen. It seemed like either Python or R would eventually disappear from the data science landscape. Hello monoculture! Some data scientists are still rooting for this, but we're guessing you're not one of them. And there was also a time when it seemed like Python and R were trying to mimic each other, just porting workflows so that language didn't matter. Luckily those endeavors have not come to fruition. Both Python and R have unique strengths; trying to imitate each other seems to miss that point.

Today many data scientists in the Python and R communities recognize that both languages are outstanding, useful, and complementary. To return to a key point in the preface, the data science community has converged onto a point of cooperation and community-building—to the benefit of everyone involved.

We're ready for a new community of bilingual data scientists. The challenge is that many users of one language don't quite know *how* they are complementary or *when* to use which language. There have been a few solutions over the years, but we'll get into that in Part IV.

Final Thoughts

At this point you should have a good idea of where we are in 2021 and how we got here. In the next part we'll introduce each group of users to a new language.

One last note: Python users refer to themselves as Pythonistias, which is a really cool name! There's no real equivalent in R, and they also don't get a really cool animal, but that's life when you're a single-letter language. R users are typically called...wait for it...useRs! (Exclamation optional.) Indeed, the official annual conference is called useR! (exclamation obligatory), and the publisher Springer has an ongoing and very excellent series of books of the same name. We'll use these names from now own.

Figure 1-1 provides a summary of some of the major events that we've highlighted in this chapter, plus some other milestones of interest.

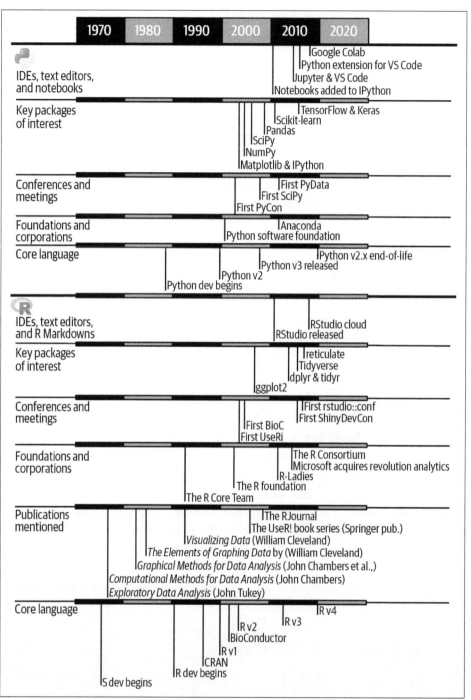

Figure 1-1. A timeline of Python and R data science milestones

## PART II

Bilingualism I: Learning a New Language

In this part, I'll introduce the two core languages for data science: Python and R. In contrast to other introductions, I expect some familiarity in one language before introducing the other. In short, I expect that you're carrying baggage. I'd like to advise you to leave your baggage at the door, but baggage is designed to be hauled around, so that's kind of hard to do. Instead, let's embrace your baggage! Recognize that Python and R operate quite differently and you may not always find a 1:1 translation. That's OK!

When I teach R and Python for complete beginners, each lesson is an *element*, a fundamental component of the whole. The first four elements are:

Functions
    How to perform actions, i.e., the verbs.

Objects
    How to store information, i.e., the nouns.

Logical Expressions
    How to ask questions.

Indexing
    How to find information.

There are many layers beyond these four elements, but these are the core, essential ones. Once you have a good grasp of these elements, you have the tools to delve further on your own. My goal is to get you to that point. Thus, the following chapters are not thorough introductions to each language.

The Appendix contains a quick-reference Python:R bilingual dictionary. It will help you translate code you're familiar with into your new, still unfamiliar language.

Chapter 2

   Begin here if you're a Pythonista who wants to get into the ueR's mindset.

Chapter 3

   Begin here if you're a useR who wants to get into the Pythonista's mindset.

Once you're familiar with your new language, continue on to Part III to learn when each is most appropriate.

R for Pythonistas

Rick J. Scavetta

Welcome, brave Pythonista, to the world of the useR![1] In this chapter I introduce you to R's core features and try to address some of the confusing bits that you'll encounter along the way. Thus, it's useful to mention what we're *not* going to do.

First, we're not writing for the naïve data scientist. If you want to learn R from scratch, there are many wonderful resources available—too many to name. We encourage you to explore them and choose those that suit your needs and learning style. Here, we'll bring up topics and concerns that may confuse the complete novice. We'll take some detours to explain topics that we hope will specifically help the friendly Pythonista to adapt to R more easily.

Second, this is not a bilingual dictionary; you'll find that in the Appendix, but without context it's not really useful. Here, we want to take you through a journey of exploRation and undeRstanding. We want you to get a *feel* for R so that you begin to *think* R, becoming bilingual. Thus, for the sake of narrative, we may introduce some items much later than when writing for a complete novice. Nonetheless, we hope that you'll return back to this chapter when you need to remind yourself of how to do familiar tasks in a new language.

Third, this is not a comprehensive guide. Once you crack the R coconut, you'll get plenty of enjoyment exploring the language more deeply to address your specific needs as they arise. As we mentioned in the first part of the book, the R community is diverse, friendly, welcoming—and helpful! We're convinced it's one of the less "tech-bro" cultures out there. To get an idea of the community, you can follow #rstats on Twitter (*https://oreil.ly/YOfeZ*).

---

1 *useR!* is the annual R conference and also a series of books by publisher Springer.

Up and Running with R

To follow the exercises in this chapter, you can either access R online using RStudio Cloud or install R and RStudio locally. RStudio Cloud is a platform providing access to an R instance (via an RStudio IDE) that allows you to upload your own data and share projects. We'll cover both methods in the following paragraphs.

To use RStudio Cloud, make an account (*https://rstudio.cloud*) and then navigate to our publically available project (*https://oreil.ly/21Sr2*). Make sure to save a copy of the project in your workspace so you have your own copy; you'll see the link in the header.

Your RStudio session should look like Figure 2-1. Open *ch02-r4py/r4py.R* and that's it! You're ready to follow along with all the examples. To execute commands, press Ctrl + Enter (or Command-Enter).

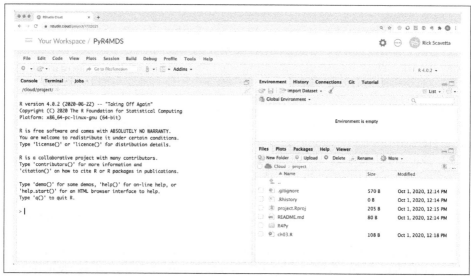

Figure 2-1. Our project in RStudio Cloud

To run R locally, you'll find it's available with the Anaconda distribution, if you use that; otherwise you can install it directly. First, download and install R for your operating system (*https://www.r-project.org*). R v4.0 was released in June 2020 and, in contrast to Python v3.x, is backward compatible, with a few notable exceptions. We'll assume you're running at least R 4.0.0: "Taking Off Again." Each release gets a name inspired by Peanuts (the classic comic strip and film franchise featuring Charlie Brown, Snoopy, and co.), which is a nice personal touch, I think. Next, install the RStudio Desktop IDE (*https://rstudio.com*).

Finally, set up a project to work on. This is a bit different from a virtual environment, which we'll discuss later on. There are two typical ways to make a project with preexisting files.

First, if you're using Git, you'll be happy to know that RStudio is also a basic Git GUI client. In RStudio, select File > "New project" > Version Control > Git and enter the repository URL *https://github.com/moderndatadesign/PyR4MDS*. The project directory name will use the repo name automatically. Choose where you want to store the repo and click Create Project.

Second, if you're not using Git, you can just download and unzip the repo from *https://github.com/moderndatadesign/PyR4MDS*. In RStudio, select File > Existing Directory and navigate to the downloaded directory. A new R project file, **.Rproj* will be created in that directory.

Your RStudio session should look like Figure 2-2. Open *ch02-r4py/r4py.R* and that's it! You're ready to follow along with all the examples. To execute commands, press Ctrl + Enter (or Command-Enter).

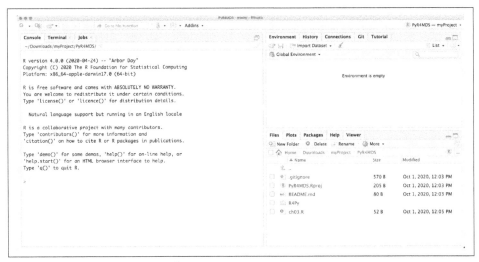

Figure 2-2. Our project in RStudio

Projects and Packages

We could begin exploring R by using a built-in dataset and diving right into the Tidyverse (introduced in Chapter 1), but I want to step back for a second, take a deep breath, and begin our story at the beginning. Let's begin by reading in a simple CSV file. For this, we're going to use a dataset that is actually already available in R in the ggplot2 package. For our purposes, we're less bothered with the actual analysis than how it's being done in R. I've provided the dataset as a file in the book repository (*https://github.com/moderndatadesign/PyR4MDS*).

If you set up your project correctly, all you'll need to execute is the following command. If this command doesn't work, don't worry, we'll return to it shortly.

```
diamonds <- read.csv("ch02-r4py/data/diamonds.csv")
```

Just like in Python, single (' ') and double ("") quotation marks are interchangeable, although there is a preference for double quotation marks.

You should now have the file imported and available as an object in your global environment, where your user-defined objects are found. The first thing you'll notice is that the environment pane of RStudio will display the object and already give some summary information. This lovely, simple touch is similar to the Jupyter Notebook extension for VS Code (see Chapter 3), which also lets you view your environment. Although this is a standard feature in RStudio, viewing a list of objects when scripting in Python, or many languages for that matter, is not typical. Clicking the little blue arrow beside the object name will reveal a text description (see Figure 2-3).

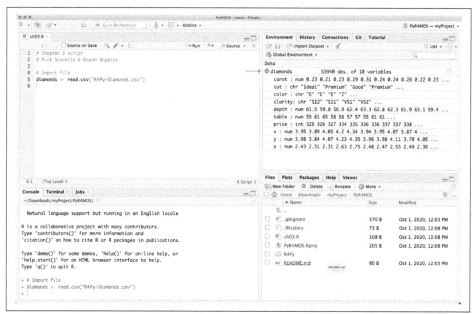

Figure 2-3. A pulldown of a data frame

Clicking on the name will open it up in an Excel-like viewer (see Figure 2-4).

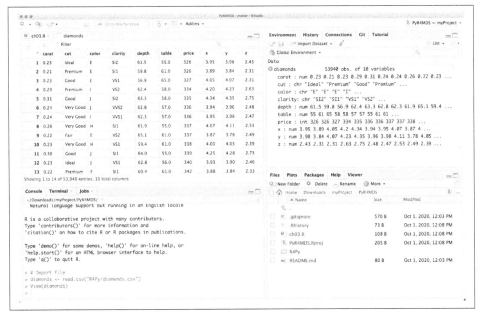

Figure 2-4. A data frame in table view

The RStudio Viewer

The RStudio viewer is much nicer than Excel, since it only loads into memory what you're seeing on the screen. You can search for specific text and filter your data here, so it's a handy tool for getting a peek at your data.

Although these are nice features, some useRs consider them to be a bit too much GUI and a bit too little IDE. Pythonistas would mostly agree, and some criticize the user experience of RStudio because of this. I partly agree, since I've seen how it can encourage bad practices. For example, to import your dataset, you could have also clicked Import Dataset. This can be convenient if you're having a really hard time parsing through the file's structure, but it leads to undocumented, nonreproducible actions that are extremely frustrating since scripts/projects will not be self-contained. The command to import the file will be executed in the console, and visible in the history panel, but it will *not* appear in the script unless you explicitly copy it. This results in objects in the environment that are not defined in the script. However, remember that RStudio is not R. You can use R with other text editors (for example the ESS [Emacs Speaks Statistics (*https://ess.r-project.org*)] extension for Emacs).

If you couldn't import your data with the previous commands, either (i) the file doesn't exist in that directory or (ii) you're working in the wrong *working directory*, which is more likely. You may be tempted to write something terrible, like this:

```
diamonds <- read.csv("ch02-r4py/data/diamonds.csv")
```

You'll be familiar with avoiding the use of hardcoded paths when using virtual environments with Python. Using relative paths, as we did earlier, ensures that our file directory contains all necessary data files. Neither the working directory nor the project are virtual environments, but they are nonetheless very handy, so let's check them out!

The working directory is the first place R looks for a file. When you use R projects, the working directory is wherever you have the *.Rproj* file. Thus, *ch02-r4py* is a sub-directory in our working directory. It doesn't matter what the working directory is called or where it is. You can move the entire project anywhere on your computer and it will still *just work* once you open the project (the *.Rproj* file) in RStudio.

 If you're not using R projects, then your working directory will likely be your home directory, displayed as *project: (None)* in RStudio. This is terrible because you'll have to specify the entire path to your file instead of just the subdirectories within your project. You'll find the command getwd() to *get*, and setwd() to *set* the working directory in many outdated tutorials. Please don't use these commands! They result in the same problems of hardcoding full file paths.

Returning to our diamonds <- read.csv("ch02-r4py/data/diamonds.csv") command, you'll already notice some things that will confuse and/or aggravate the seasoned Pythonista. Three things in particular stand out.

First, notice that it's commonplace, and even preferred, to use <- as the assign operator in R. You can use =, as in Python, and indeed you'll see prominent and experienced useRs do this, but <- is more explicit as *assign to object* since = is also used to assign values to arguments in function calls, and we all know how much Pythonistas love being explicit!

 The <- assign operator is actually a legacy operator stemming from the prestandardized QWERTY keyboard where the <- didn't mean *move the cursor one space to the left* but literally, *make <- appear.*

Second, notice that the function name is read.csv(). Nope, that's not a typo. csv() is not a *method* of object read, nor is it a *function* of module read. Both are completely acceptable interpretations if this was a Python command. In R, with a few, but notable, exceptions, "." doesn't mean anything special. It's a bit annoying if you're used to more object-oriented programming (OOP) languages where "." is a special character.

Finally, you'll notice that we didn't initialize any packages to accomplish this task. The `read.*()` function variants are a part of base R. Interestingly, there are newer and more convenient ways of reading in files if these functions don't satisfy your needs. For example, the `read_csv()` function is in the readr package. We know you're excited to see that "_"!

In general, when you see simple functions with ".," these are old base R functions created when nobody worried that it would be confusing to have "." in the names. Functions from the newer Tidyverse packages (e.g., readr) tend to use "_" (see Chapter 1). They basically do the same thing, but with some slight tweaks to make them more user friendly.

Let's see this in action with readr. Just like in Python, you'll need to install the package. This is typically done directly in the R console; there is no `pip` equivalent in R.

Use the following command:

```
install.packages("tidyverse")
```

 In RStudio, you can install packages by using the Packages panel in the lower-right pane and clicking Install. Type in **tidyverse** and make sure that the Install All Dependencies box is checked and click OK. If you go this route, refrain from clicking on the checkboxes beside the names of the installed packages. This will initialize the package, but not record it in your script.

This will by default install packages and their dependencies from CRAN, the repository of official R packages. Official packages have undergone quality control and are hosted on mirrored servers around the world. The first time you do this, you'll be asked to choose a mirror site to install from. For the most part it doesn't matter which one you choose. You'll see a lot of red text as the core Tidyverse packages and all their dependencies are installed. This is mostly just a convenient way to get lots of useful packages installed all at once.

The most common problem in installing packages is to not have write permission in the packages directory. This will prompt you to create a personal library. You can always check where your packages are installed by using

```
.libPaths()
[1] "/Library/Frameworks/R.framework/Versions/4.0/Resources/library"
```

If you have a personal library, it will be shown here in the second position.

In contrast to Pythonistas, who tend to use virtual environments, useRs typically install a package once, making it available system-wide. After many false starts in trying to implement a solution for project-specific libraries in R, the current favorite is the renv package (*https://oreil.ly/GjXEw*), i.e., *R environment*.

As in Python, after installing a package, it needs to be initialized in each new R session. When we say *initialize*, or *load*, a package, what we're really saying is "use the `library()` function to *load* an installed package and then *attach* it to the namespace, i.e., the global environment." All your packages comprise your *library*, hence `library()`. The core suite of packages in the Tidyverse can be loaded using `library(tidyverse)`. That is commonplace and for the most part not a problem, but you may want to get into the habit of loading only those packages that you actually require instead of filling up your environment needlessly. Let's start with readr, which contains the `read_csv()` function.

```
# R
library(readr)
```

This is the equivalent of:

```
# Python equivalent
import readr
```

Although R uses OOP, it's mostly operating in the background, hence you'll never see strange aliases for packages like:

```
import readr as rr
```

That's just a foreign concept in R. After you have *attached* the package, all functions and datasets in that package are available in your global environment.

 This calls to mind another legacy function that you may see floating around. You must absolutely avoid `attach()` (and for the most part its counterpart `detach()`). This function allows you to *attach* an object to your global environment, much like how we attached a package. Thus, you can call elements within the object directly, without first specifying the object name explicitly, like how we call functions within a package without having to explicitly call the package name every time. The reason this has fallen out of favor is that you're likely to have many data objects that you want to access, so conflicting names are likely to be an issue (i.e., leading to *masking* of objects). Plus, it's just not explicit.

Let's address another issue with loading packages before continuing: you'll often see:

```
require(readr)
```

`require()` will load an installed package and also return a TRUE/FALSE based on success. This is useful for testing if a package exists, so it should be reserved for those instances where that is necessary. For the most part, you want to use `library()`.

Alright, let's read in our dataset again, this time using `read_csv()` to make some simple comparisons between the two methods.

```
> diamonds_2 <- read_csv("R4Py/diamonds.csv")
Parsed with column specification:
cols(
  carat = col_double(),
  cut = col_character(),
  color = col_character(),
  clarity = col_character(),
  depth = col_double(),
  table = col_double(),
  price = col_double(),
  x = col_double(),
  y = col_double(),
  z = col_double()
)
```

You'll notice that we're afforded a more detailed account of what's happened.

As we mentioned earlier, Tidyverse design choices tend to be more user friendly than older processes they update. This output tells us the column names of our tabular data and their types (see Table 2-2).

Also note that the current trend in R is to use *snake case*, underscores ("_") between words and only lowercase letters. Although there has classically been poor adherence to a style guide in R, *Advanced R* (*https://oreil.ly/PmoJl*) by Hadley Wickham (CRC Press) offers good suggestions. Google also attempted to promote an R style guide (*https://oreil.ly/24XzZ*), but it doesn't seem that the community is very strict on this issue. This is in contrast to a strict adherence to the PEP 8 Style Guide for Python Code, authored by Guido van Rossum and released in the early days of Python.

The Triumph of Tibbles

So far, we've imported our data twice, using two different commands. This was done so that you can see some of how R works under the hood and some typical behavior of the Tidyverse versus the base package. We already mentioned that you can click on the object in the Environment Viewer, the upper-right pane, to look at it, but it's also typical to just print it to the console. You may be tempted to execute:

```
> print(diamonds)
```

But the print() function is not necessary except in specific cases, like within a for loop. As with a Jupyter Notebook, you can just execute the object name; for example:

```
> diamonds
```

This will print the object to the console. We won't reproduce it here, but if you do execute the > diamonds command, you'll notice that this is not a nice output! Indeed, one wonders why the default output allows so much to be printed to the console in interactive mode. Now try with the data frame we read in using read_csv():

```
> diamonds_2
# A tibble: 53,940 x 10
   carat cut       color clarity depth table price     x     y     z
   <dbl> <chr>     <chr> <chr>   <dbl> <dbl> <dbl> <dbl> <dbl> <dbl>
 1  0.23 Ideal     E     SI2      61.5    55   326  3.95  3.98  2.43
 2  0.21 Premium   E     SI1      59.8    61   326  3.89  3.84  2.31
 3  0.23 Good      E     VS1      56.9    65   327  4.05  4.07  2.31
 4  0.290 Premium  I     VS2      62.4    58   334  4.2   4.23  2.63
 5  0.31 Good      J     SI2      63.3    58   335  4.34  4.35  2.75
 6  0.24 Very Good J     VVS2     62.8    57   336  3.94  3.96  2.48
 7  0.24 Very Good I     VVS1     62.3    57   336  3.95  3.98  2.47
 8  0.26 Very Good H     SI1      61.9    55   337  4.07  4.11  2.53
 9  0.22 Fair      E     VS2      65.1    61   337  3.87  3.78  2.49
10  0.23 Very Good H     VS1      59.4    61   338  4     4.05  2.39
# … with 53,930 more rows
```

Wow! That's a much nicer output than the default base R version. We have a neat little table with the names of the columns on one row, and three-letter codes for the data types under them that are set in angle brackets (<>). We only see the first 10 rows and then a note telling us how much we're not seeing. If there were too many columns for our screen, we'd see them listed at the bottom. Give that a try; set your console output to be very narrow and execute the command again:

```
# A tibble: 53,940 x 10
   carat cut     color clarity
   <dbl> <chr>   <chr> <chr>
 1  0.23 Ideal   E     SI2
 2  0.21 Premium E     SI1
 3  0.23 Good    E     VS1
 4  0.290 Premium I    VS2
 5  0.31 Good    J     SI2
 6  0.24 Very G… J     VVS2
 7  0.24 Very G… I     VVS1
 8  0.26 Very G… H     SI1
 9  0.22 Fair    E     VS2
10  0.23 Very G… H     VS1
# … with 53,930 more rows,
#   and 6 more variables:
#   depth <dbl>, table <dbl>,
#   price <dbl>, x <dbl>,
#   y <dbl>, z <dbl>
```

Base R was already pretty good for exploratory data analysis (EDA), but this is next-level convenience. So what happened? Actually understanding this is pretty important, but first we want to highlight two other interesting points.

First, notice that we didn't need to load all of readr to gain access to the read_csv() function. We could have left out library(readr) and just used:

```
> diamonds_2 <- readr::read_csv("R4Py/diamonds.csv")
```

The double-colon operator :: is used to access functions within a package, akin to:

```
from pandas import read_csv
```

You'll see :: used when useRs know that they'll only need one very specific function from a package, or that functions in two packages may conflict with each other, so they want to avoid attaching an entire package to their namespace.

Second, this is the first time we see actual data in R, and we can tell right away that numbering begins with 1! (And why wouldn't it?).

Printing Objects to the Screen

Just as an aside for printing objects to the screen, you'll often see round brackets around an entire expression. This just means to execute the expression and print the object to the screen:

```
(aa <- 8)
```

It mostly just clutters up commands. Unless it's necessary, just explicitly call the object:

```
aa <- 8
aa
```

Plus, it's easier to just comment out (use Ctrl + Shift + C in RStudio) the print line instead of having to go back and remove all those extra brackets.

OK, let's get to the heart of what's happening. Why do diamonds and diamonds_2 *look* so different when printed to the console? Answering this question will help us to understand a bit about how R handles objects. To answer this question, let's take a look at the class of these objects:

```
class(diamonds)
[1] "data.frame"

class(diamonds_2)
[1] "spec_tbl_df" "tbl_df"      "tbl"         "data.frame"
```

You'll be familiar with a data.frame from pd.DataFrame (OK, can we just admit that a pd.DataFrame is just a Python implementation of an R data.frame?). But using the

Tidyverse `read_csv()` function produced an object with three additional classes. The two to mention here are the subclass `tbl_df` and the class `tbl`; the two go hand in hand for defining a *tibble* (hence `tbl`), which has a data frame structure `tbl_df`.

Tibbles are a core feature of the Tidyverse and have many perks over base R objects. For example, printing to the console. Recall that calling an object name is just a shortcut for calling `print()`, which has a method to handle data frames. Now that we've attached the readr package, it has a method to handle objects of class `tbl_df`.

So here we see OOP principles operating in the background implicitly handling object classes and calling the methods appropriate to a given class. Convenient! Confusing? Implicit! I can see why Pythonistas get annoyed, but once you get over it, you see that you can just get on with your work without too much hassle.

A Word About Types and Exploring

Let's take a deeper look at our data and see how R stores and handles data. A data frame is a two-dimensional heterogeneous data structure. It sounds simple, but let's break it down a bit further (see Table 2-1).

Table 2-1. Common data structures in R

Name	Number of dimensions	Type of data
Vector	1	Homogeneous
List	1	Heterogeneous
Data frame	2	Heterogeneous
Matrix	2	Homogeneous
Array	n	Homogeneous

Vectors are the most basic form of data storage. They are one-dimensional and homogeneous. That is, one element after another, where every element is of the same type. It's like a one-dimensional NumPy array composed solely of scalars. We don't refer to scalars in R; that's just a one-element-long vector. There are many *types* in R, and four commonly used "user-defined atomic vector types." The term *atomic* already tells us that it doesn't get any more basic than what we find in Table 2-2.

Table 2-2. The four most common user-defined atomic vector types in R

Type	Data frame shorthand	Tibble shorthand	Description
Logical	logi	<lgl>	Binary TRUE/FALSE, T/F, 1/0
Integer	int	<int>	Whole numbers from [-Inf,Inf]
Double	num	<dbl>	Real numbers from [-Inf,Inf]
Character	chr	<chr>	All alpha-numeric characters, including white spaces

The two other, less common, user-defined atomic vector types are `raw` and `complex`.

Vectors are fundamental building blocks. There are a few things to note about vectors, so let's get that out of the way before we return to the workhorse of data science, the beloved data frame.

The four user-defined atomic vector types listed in Table 2-2 are ordered according to increasing levels of information content. When you create a vector, R will try to find the lowest information-content type that can encompass all the information in that vector. For example, `logical`:

```
> a <- c(TRUE, FALSE)
> typeof(a)
[1] "logical"
```

`logical` is R's equivalent of `bool`, but it is very rarely referred to as *boolean* or *binary*. Also, note that `T` and `F` are not in themselves reserved terms in R, so they are not recommended for logical vectors, although they are valid. Use `TRUE` and `FALSE` instead. Let's take a look at numbers:

```
> b <- c(1, 2)
> typeof(b)
[1] "double"

> c <- c(3.14, 6.8)
> typeof(c)
[1] "double"
```

R will automatically convert between `double` and `integer` as needed. Math is performed primarily using double-precision, which is reflected in the data frame shorthand for `double` being displayed as `numeric`. Unless you explicitly need to restrict a number to be a true integer, then `numeric`/`double` will be fine. If you do want to restrict values to be `integer`, you can *coerce* them to a specific type using one of the `as.*()` functions, or use the `L` suffix to specify that a number must be an integer.

```
> b <- as.integer(c(1, 2))
> typeof(b)
[1] "integer"

> b <- c(1L, 2L)
> typeof(b)
[1] "integer"
```

Characters are R's version of strings. You'll know this as `str` in Python, which is, confusingly, a common R function, `str()`, which gives the *structure* of an object. Characters are also frequently referred to as strings in R, including in arguments and package names, which is an unfortunate inconsistency:

```
> d <- c("a", "b")
> typeof(d)
[1] "character"
```

Putting these together in a vanilla data frame using `data.frame()` or using the more recently developed tibble using `tibble()` gives us:

```
my_tibble <- tibble(a = c(T, F),
                    b = c(1L, 2L),
                    c = c(3.14, 6.8),
                    d = c("a", "b"))
my_tibble

# A tibble: 2 x 4
  a     b     c d
  <lgl> <int> <dbl> <chr>
1 TRUE      1  3.14 a
2 FALSE     2  6.8  b
```

Notice we get the nice output from `print()` since it's a tibble. When we look at the *structure*, we'll see some confusing features:

```
> str(my_tibble)
tibble [2 × 4] (S3: tbl_df/tbl/data.frame)
 $ a: logi [1:2] TRUE FALSE
 $ b: int [1:2] 1 2
 $ c: num [1:2] 3.14 6.8
 $ d: chr [1:2] "a" "b"
```

`str()` is a classic base-package function and gives some bare-bones output. It's similar to what you'll see when you click on the reveal arrow beside the object's name in the environment panel. The first row gives the object's class (which we already saw). S3 refers to the specific OOP system that this object uses, which in this case is the most basic and least strict OOP system.

Alternatively, we can also use the Tidyverse `glimpse()` function, from the dplyr package:

```
> library(dplyr)
> glimpse(my_tibble)
Rows: 2
Columns: 4
$ a <lgl> TRUE, FALSE
$ b <int> 1, 2
$ c <dbl> 3.14, 6.80
$ d <chr> "a", "b"
```

Notice that Table 2-2 also states the shorthand `num`, which does not appear in the output of `glimpse()`. This refers to the the "numeric" class, indicating either `double` (for double-precision floating-point numbers) or `integer` type.

The preceding examples showed us that a `data.frame` is a heterogeneous, two-dimensional collection of homogeneous one-dimensional vectors, each having the same length. We'll get to why R prints all those dollar signs (and no, it has nothing to do with your salary!).

Naming (Internal) Things

We already mentioned that snake case is the current trend in naming objects in R. However, naming columns in a data frame is a different beast altogether because we just inherit names from the first line of the source file. Data frames in base R, obtained for example, using the `read.*()` family of functions or manually created using the `data.frame()` function, don't allow for any *illegal* characters. Illegal characters include all white spaces and all reserved characters in R:

- Arithmetic operators (+, -, /, *, etc.)
- Logical operators (&, |, etc.)
- Relational operators (==, !=, >, <, etc.)
- Brackets ([, (, {, <, and their closers)

In addition, although they can *contain* numbers, they can't *begin* with numbers. Let's see what happens:

```
# Base package version
data.frame("Weight (g)" = 15,
           "Group" = "trt1",
           "5-day check" = TRUE)
  Weight..g. Group X5.day.check
1         15  trt1         TRUE
```

All the illegal characters have been replaced with .! I know, right? R is *really* having a good time mocking you OOP obsessives! On top of that, any variable that began with a number is now prefaced with an X.

So what about importing a file with no header?

```
> diamonds_base_nohead <- read.csv("ch02-r4py/data/diamonds_noheader.csv",
                                    header = F)
> names(diamonds_base_nohead)
 [1] "V1"  "V2"  "V3"  "V4"  "V5"  "V6"  "V7"  "V8"  "V9"  "V10"
```

In base R, if we don't have any header, the given names are V for "variable" followed by the number of that column.

The same file read in with one of the `readr::read_*()` family of functions or created with `tibble()` will maintain illegal characters! This seems trivial, but it's actually a serious critique of the Tidyverse, and it's something to pay close attention to if you start meddling in other people's scripts. Let's look:

```
> tibble("Weight (g)" = 15,
+        "Group" = "trt1",
+        "5-day check" = TRUE)
# A tibble: 1 x 3
  `Weight (g)` Group `5-day check`
```

```
          <dbl> <chr> <lgl>
1            15 trt1  TRUE
```

Notice the paired backticks for the column Weight (g) and 5-day check? You now need to use this to escape the illegal characters. Perhaps this makes for more informative commands, since you have the full name, but you'll likely want to maintain short and informative column names anyways. Information about the unit (e.g., *g* for weight) is extraneous information that belongs in a dataset legend.

Not only that, but the names given to datasets without headers are also different:

```
> diamonds_tidy_nohead <- read_csv("ch02-r4py/data/diamonds_noheader.csv",
                          col_names = F)
> names(diamonds_tidy_nohead)
 [1] "X1"  "X2"  "X3"  "X4"  "X5"  "X6"  "X7"  "X8"  "X9"  "X10"
```

Instead of V we get X! This takes us back to the Tidyverse as a distinct dialect in R. If you inherit a script entirely in base R, you'll have a tricky time if you just start throwing in Tidyverse functions with wild abandon. It's like asking for a Berliner in a Berlin bakery![2]

Lists

Lists are another common data structure, but they're not exactly what you expect from a Python list, so the naming can be confusing. Actually, we've already encountered lists in our very short R journey. That's because the data.frame is a specific class of type list. Yup, you heard that right.

```
> typeof(my_tibble)
[1] "list"
```

Table 2-1 tells us that a list is a one-dimensional, heterogeneous object. What that means is that every element in this one-dimensional object can be a different type. Indeed lists can contain not only vectors but other lists, data frames, matrices, and on and on. In the case that each element is a vector of the same length, we end up with tabular data that is then class data.frame. Pretty convenient, right? Typically, you'll encounter lists as the output from statistical tests; let's take a look.

The PlantGrowth data frame is a built-in object in R. It contains two variables (i.e., elements in the list, aka columns in the tabular data): weight and group.

```
> glimpse(PlantGrowth)
Rows: 30
Columns: 2
$ weight <dbl> 4.17, 5.58, 5.18, 6.11, 4.50, 4.6...
$ group  <fct> ctrl, ctrl, ctrl, ctrl, ctrl, ctr...
```

---

2 Berliner (noun): in Berlin, a resident of the city. Everywhere else: a tasty jelly-filled, sugar-powered donut.

The dataset describes the dry plant `weight` (in grams; thank you, data legend) of 30 observations (i.e., individual plants, aka rows in the tabular data) grown under one of three conditions described in `groups`: `ctrl`, `trt1`, and `trt2`. The convenient `glimpse()` function doesn't show us these three groups, but the classic `str()` does:

```
> str(PlantGrowth)
'data.frame':   30 obs. of  2 variables:
 $ weight: num  4.17 5.58 5.18 6.11 4.5 4.61 5.17 4.53 5.33 5.14...
 $ group : Factor w/ 3 levels "ctrl","trt1",..: 1 1 1 1 1 1 1 1 1 1...
```

If you're getting nervous about `<fct>` and `Factor w/ 3 levels`, just hang tight—we'll talk about that after we're done with lists.

Alright, let's get to some tests. We may want to define a linear model for `weight` described by `group`:

```
pg_lm <- lm(weight ~ group, data = PlantGrowth)
```

`lm()` is a foundational and flexible function for defining linear models in R. Our model is written in *formula notation*, where `weight ~ group` is y ~ x. You'll recognize the ~ as the standard symbol for "described by" in statistics. The output is a type `list` of class `lm`:

```
> typeof(pg_lm)
[1] "list"
> class(pg_lm)
[1] "lm"
```

There are two things that we want to remind you of and build on here.

First, remember that we mentioned that a data frame is a collection of vectors of the same length? Now we see that that just means that it's a special class of a type `list`, where each *element* is a vector of the same length. We can access a named element within a list using the $ notation:

```
> names(PlantGrowth)
[1] "weight" "group"
> PlantGrowth$weight
 [1] 4.17 5.58 5.18 6.11 4.50 4.61 5.17 4.53 5.33 5.14 4.81 4.17 4.41 3.59
[15] 5.87 3.83 6.03 4.89 4.32 4.69 6.31 5.12 5.54 5.50 5.37 5.29 4.92 6.15
[29] 5.80 5.26
```

Notice the way it's printed, along a row, and the beginning of each row begins with a [] with an index position in there. (We already mentioned that R begins indexing at 1, right?) In RStudio, you'll get an autocomplete list of column names after typing **$**.

We can also access a named element within a list using the same notation:

```
> names(pg_lm)
[1] "coefficients"  "residuals"     "effects"       "rank"
[5] "fitted.values" "assign"        "qr"            "df.residual"
```

```
 [9] "contrasts"     "xlevels"       "call"          "terms"
[13] "model"
```

You can see how a list is such a nice way to store the results of a statistical test since we have lots of different kinds of output. For example, `coefficients`:

```
> pg_lm$coefficients
(Intercept)    grouptrt1    grouptrt2
      5.032       -0.371        0.494
```

is a *named* three-element-long numeric vector. (Although its elements are named, the $ operator is invalid for atomic vectors, but we have some other tricks up our sleeve, of course—see indexing with [ ] in "How to Find...Stuff" on page 32.) We didn't get into the details, but you may be aware that given our data we expect to have three coefficients (estimates) in our model.

Consider `residuals`:

```
> pg_lm$residuals
      1      2      3      4      5      6      7      8      9     10
 -0.862  0.548  0.148  1.078 -0.532 -0.422  0.138 -0.502  0.298  0.108
     11     12     13     14     15     16     17     18     19     20
  0.149 -0.491 -0.251 -1.071  1.209 -0.831  1.369  0.229 -0.341  0.029
     21     22     23     24     25     26     27     28     29     30
  0.784 -0.406  0.014 -0.026 -0.156 -0.236 -0.606  0.624  0.274 -0.266
```

They are stored in a named 30-element-long numerical vector (remember we had 30 observations). So lists are pretty convenient for storing heterogeneous data and you'll see them quite often in R, although there is a concerted effort in the Tidyverse toward data frames and their variants thereof.

Second, remember we mentioned that the . mostly doesn't have any special meaning. Well here's one of the exceptions where the . does actually have a meaning. Probably the most common use is when it specifies *all* when defining a model. Here, since other than the `weight` column, `PlantGrowth` only had one other column, we could have written:

```
lm(weight ~ ., data = PlantGrowth)
```

It's not really necessary, since we only have one independent variable here, but in some cases it's convenient. The `ToothGrowth` dataset has a similar experimental setup, but we're measuring the length of tooth growth under two conditions: a specific supplement (`supp`) and its dosage (`dose`).

 A note on variable types. By using the y ~ x formula, we're say that x is the independent or predictor variable(s) and y is dependent on x, or the response to the predictor.

```
lm(len ~ ., data = ToothGrowth)
# is the same as
lm(len ~ supp + dose, data = ToothGrowth)
```

But like always, being explicit has its advantages, such as defining more precise models:

```
lm(len ~ supp * dose, data = ToothGrowth)
```

Can you spot the difference between the two outputs? Specifying interactions is done with the *.[3]

The Facts About Factors

Alright, the last thing we need to clear up before we continue is the phenomena of the factor. Factors are akin to the pandas category type in Python. They are a wonderful and useful class in R. For the most part they exist and you won't have cause to worry about them, but do be aware, since their uses and misuses will make your life a dream or a misery, respectively. Let's take a look.

The name "factor" is very much a statistics term. We may refer to them as categorical variables as Python does, but you'll also see them referred to as qualitative and discrete variables, in textbooks and also in specific R packages, like RColorBrewer and ggplot2, respectively. Although these terms all refer to the same *kind* of variable, when we say "factor" in R, we're referring to a *class of type integer*. It's like how data.frame is a *class of type list*. Observe:

```
> typeof(PlantGrowth$group)
[1] "integer"
> class(PlantGrowth$group)
[1] "factor"
```

You can easily identify a factor because in both the output from str() (see "Lists" on page 28) and in plain vector formatting, the levels will be stated:

```
> PlantGrowth$group
 [1] ctrl ctrl ctrl ctrl ctrl ctrl ctrl ctrl ctrl ctrl
[11] trt1 trt1 trt1 trt1 trt1 trt1 trt1 trt1 trt1 trt1
[21] trt2 trt2 trt2 trt2 trt2 trt2 trt2 trt2 trt2 trt2
Levels: ctrl trt1 trt2
```

The levels are statisticians' names for what are commonly called *groups*. Another giveaway is that, although we have characters, they are not enclosed in quotation marks! This is very curious because we can actually treat them as characters, even though they are type integer (see Table 2-2). You may be interested to look at the

---

3 A detailed exposition of model definitions is outside the scope of this text.

internal structure of an object using dput(). Here we can see that we have an integer vector c(1L...) and two *attributes*, the label and the class:

```
> dput(PlantGrowth$group)
structure(c(1L, 1L, 1L, 1L, 1L, 1L, 1L, 1L, 1L, 1L,
            2L, 2L, 2L, 2L, 2L, 2L, 2L, 2L, 2L, 2L,
            3L, 3L, 3L, 3L, 3L, 3L, 3L, 3L, 3L, 3L),
          .Label = c("ctrl", "trt1", "trt2"),
          class = "factor")
```

The labels define the names of each *level* in the factor and are mapped to the integers, 1 being ctrl, and so on. So when we print to the screen we only see the names, not the integers. This is commonly accepted to be a legacy use case from the days when memory was expensive and it made sense to save an integer many times over instead of a potentially long character vector.

So far, the only kind of factor we've seen was really describing a nominal variable (a categorical variable with no order), but we have a nice solution for ordinal variables also. Check out this variable from the diamonds dataset:

```
> diamonds$color
[1] E E E I J J I H E H ..
Levels: D < E < F < G < H < I < J
```

The levels have an order, in the sense that D comes before E, and so on.

How to Find...Stuff

Alright, by now we've seen how R stores data and various subtleties that you'll need to keep in mind, in particular things that may trip up a Pythonista. Let's move on to logical expressions and indexing, which is to say: how to find...stuff?

Logical expressions are combinations of relational operators, which *ask* yes/no questions of *comparison*, and logical operators, which *combine* those yes/no questions.

Let's begin with a vector:

```
> diamonds$price > 18000
    [1] FALSE FALSE FALSE FALSE FALSE FALSE
    ...
```

This simply asks which of our diamonds are more expensive than $18,000. There are three key things to always keep in mind here.

First, the length of the shorter object, here the unassigned numeric vector 18000 (one element long) will be "recycled" over the entire length of the longer vector, here the price column from the diamonds data frame accessed with the $ notation (53,940 elements). In Python you may refer to this as *broadcasting* when using NumPy arrays, and *vectorization* as a distinct function. In R, we simply refer to both as *vectorization* or *vector recycling*.

Second, this means that the output vector is the same length as the length of the longest vector, here 53,940 elements.

Third, anytime you see a relational or logical operator, you know that the output vector will *always* be a logical vector. (Logical as in TRUE/FALSE, not as in Mr. Spock.)

If you want to combine questions, you'll have to combine two complete questions, such as really expensive and small diamonds (classy!):

```
> diamonds$price > 18000 & diamonds$carat < 1.5
  [1] FALSE FALSE FALSE FALSE FALSE FALSE
  ...
```

Notice that all three key points hold true. When I introduced the atomic vector types, I failed to mention that logical is also defined by 1 and 0. This means we can do math on logical vectors, which is very convenient. How many expensive little diamonds do we have?

```
> sum(diamonds$price > 18000 & diamonds$carat < 1.5)
  [1] 9
```

(Not enough, if I'm being honest.) What proportion of my dataset do they represent? Just divide by the total number of observations:

```
> sum(diamonds$price > 18000 & diamonds$carat < 1.5)/nrow(diamonds)
  [1] 0.0001668521
```

So that's asking and combining questions. Let's take a look at indexing using []. You're already familiar with [], but I feel that they are more straightforward in R right out of the box. A summary is given in Table 2-3.

Table 2-3. Indexing

Use	Data object	Result
xx[i]	Vector	Vector of only i elements
xx[i]	List, Data frame, tibble	The i element maintaining the original structure
xx[[i]]	List, Data frame, tibble	The i element extracted from a list
xx[i,j]	Data frame, tibble, or matrix	The i rows and j columns of a data frame, tibble, or matrix
xx[i,j,k]	Array	The i rows, j columns, and k dimension of an array

i, j, and k are three different types of vector that can be used inside []:

- An integer vector
- A logical vector
- A character vector containing names, if the elements are named

This should be familiar to you already from Python. For integer and logical vectors, these can be unassigned vectors, or objects or functions that resolve to integer or logical vectors. Numbers don't need to be type integer, although whole numbers are clearer. Using numeric/double rounds *down* to the nearest whole number, but try to avoid using real numbers when indexing, unless it serves a purpose.

Let's begin with integers. We'll take another little detour here to discuss the omnipresent : operator, which won't do what your Pythonista brain tells you it should do. We'll begin with a built-in character vector, letters, which is the same as having a column in a data frame, like PlantGrowth$weight:

```
> letters[1] # The 1st element (indexing begins at 1)
[1] "a"
```

So that's pretty straightforward. How about counting backward?

```
> letters[-4] # Everything except the 4th element,
> # (*not* the fourth element, counting backward!)
 [1] "a" "b" "c" "e" "f" "g" "h" ...
```

Nope, that's not happening, the - means to exclude an element, *not* to count backward, but it was a nice try. We can also exclude a range of values:

```
> letters[-(1:20)] # Exclude elements 1 through 20
[1] "u" "v" "w" "x" "y" "z"
```

and of course index a range of values:

```
> letters[23:26] # The 23rd to the 26th element
[1] "w" "x" "y" "z"
```

And remember, we can combine this with anything that will give us an integer vector. length() will tell us how many elements we have in our vector, and lhs:rhs is shorthand for the function seq(from = lhs, to = rhs, by = 1), which creates a sequence of values in incremental steps of by, in this case defaulting to 1.

```
>    # The 23rd to the last element
[1] "w" "x" "y" "z"
```

So that means you always need an lhs and an rhs when using :. It's a pity, but this isn't going to work:

```
> letters[23:] # error
```

Using the [ ] inappropriately gives rise to a legendary and mysterious error message in R:

```
> df[1]
Error in df[1] : object of type 'closure' is not subsettable
> t[6]
Error in t[6] : object of type 'closure' is not subsettable
```

Can you tell where we went wrong? df and t are not defined data storage objects that we can index! They are functions, and thus they must be followed by () where we provide the arguments. [] are always used to *subset*, and these functions (df() and t()) are functions of type closure, which are *not* subsettable. So it's a pretty clear error message actually, and a good reminder to not call objects using ambiguous, short names, or indeed to get confused between functions and data storage objects.

That's all fine and good, but you're probably aware that the true power in indexing comes from using logical vectors to index specific TRUE elements, just like using type bool in Python. The most common way of obtaining a logical vector for indexing is to use a logical expression, as we've discussed. This is exactly what happens with *masking* in NumPy.

So what are the colors of those fancy diamonds?

```
> diamonds$color[diamonds$price > 18000 & diamonds$carat < 1.5]
[1] D D D D F D F F E
Levels: D < E < F < G < H < I < J
```

Here, we're using price and carat to find the colors of the diamonds that we're interested in. Not surprisingly, they are the best color classifications. You may find it annoying that you have to write diamonds$ repeatedly, but we would argue that it just makes it more explicit, and it's what happens when we reference pandas Series in Python. Since we're indexing a vector, we get a vector as output. Let's turn to data frames. We could have written the preceding indexing command as:

```
> diamonds[diamonds$price > 18000 & diamonds$carat < 1.5, "color"]
# A tibble: 9 x 1
  color
  <ord>
1 D
2 D
3 D
4 D
5 F
6 D
7 F
8 F
9 E
```

As you would expect, in [i,j], i always refers to the *rows* (observations), and j always refers to *columns* (variables). Notice that we also mixed two different types of input, but it works because they were in different parts of the expression. We use a logical vector that is as long as the data frame's number of observations (thank you, vector recycling) to obtain all the TRUE rows, and then we used a character vector to extract a named element; recall that each column in a data frame is a named element. This is a really typical formulation in R. The output is a data frame, specifically a tibble because we used indexing on the diamonds data frame and not on a specific one-dimensional vector therein. Not to get bogged down with the topic, but it is worth noting that if we didn't have a tibble, indexing for a single column (in j) would return a vector:

```
> class(diamonds)
[1] "data.frame"
> diamonds[diamonds$price > 18000 & diamonds$carat < 1.5, "color"]
[1] D D D D F D F F E
Levels: D < E < F < G < H < I < J
```

This is indeed confusing and highlights the necessity to always be aware of the class of our data object. The Tidyverse tries to address some of this by maintaining data frames even in those instances where base R prefers to revert to a vector. The Tidyverse functions for indexing, shown in the following, make things easier. (The base package shorthand, `subset()`, works much in the same way, but `filter()` works better when used in a Tidyverse context.)

```
> diamonds %>%
+   filter(price > 18000, carat < 1.5) %>%
+   select(color)
# A tibble: 9 x 1
  color
  <ord>
1 D
2 D
3 D
4 D
5 F
6 D
7 F
8 F
9 E
```

We introduced the principles behind the Tidyverse in the first part of the book, and now we're seeing it in action. The forward pipe or pipe operator, "%>%", in the preceding code allows us to *unnest* objects and functions. For example, we could have written:

```
> select(filter(diamonds, price > 18000, carat < 1.5), color)
```

That has the format of a long, nested function that is quite difficult to follow. We can pronounce %>% as "and then" and thus read the entire command above as "Take the diamonds dataset and then filter using these criteria and then select only these columns." This goes a long way in helping us to literally read and understand code and is why dplyr is described as the *grammar of data analysis*. Objects, like tibbles, are the nouns, %>% is our punctuation, and functions are the verbs.

Although Tidyverse functions are often piped together using %>%, this operator can be used to unnest nested functions. Given this widespread use, a native forward pipe operator, "|>", was included in R v4.1 (released on 18 May 2021, as we finalized this text). It is unclear how useRs will adopt it, given the dominance of %>% and some slight differences between the two operators. For now, you can safely stick with %>%.

The five most important verbs in dplyr are listed in Table 2-4.

Table 2-4. Function description

Function	Works on	Description
filter()	rows	Use a logical vector to retain only TRUE rows
arrange()	rows	Reorder rows according to values in a specific column
select()	columns	Use a name or a helper function to extract only those columns
summarise()	columns	Apply aggregation functions to a column
mutate()	columns	Apply transformation functions to a column

We already saw `filter()` and `select()` in action, so let's take a look at applying functions with `summarise()` and `mutate()`. `summarise()` is used to apply an *aggregation* function, which returns a single value like the mean, `mean()`, or standard deviation, `sd()`. It's common to see `summarise()` used in combination with the `group_by()` function. In our analogy of grammatical elements, `group_by()` is an adverb; it modifies how a verb operates. In the following example, we use `group_by()` to add a `Group` attribute to our data frame, and the functions applied in `summarise` are thus group-specific. It's just like the `.groupby()` method for `pd.DataFrames`!

```
> PlantGrowth %>%
+   group_by(group) %>%
+   summarise(avg = mean(weight),
+             stdev = sd(weight))
`summarise()` ungrouping output (override with `.groups` argument)
# A tibble: 3 x 3
  group   avg stdev
  <fct> <dbl> <dbl>
1 ctrl   5.03 0.583
2 trt1   4.66 0.794
3 trt2   5.53 0.443
```

`mutate()` is used to apply a *transformation* function, which returns as many outputs as inputs. In these cases it's not unusual to use both Tidyverse syntax and native [ ] in combination to index specific values. For example, this dataset contains the area under irrigation (thousands of hectares) for different regions of the world at four different time points:

```
> irrigation <- read_csv("R4Py/irrigation.csv")
Parsed with column specification:
cols(
  region = col_character(),
  year = col_double(),
  area = col_double()
)
> irrigation
# A tibble: 16 x 3
```

```
   region      year  area
   <chr>      <dbl> <dbl>
1 Africa      1980   9.3
2 Africa      1990  11
3 Africa      2000  13.2
4 Africa      2007  13.6
5 Europe      1980  18.8
6 Europe      1990  25.3
7 Europe      2000  26.7
8 Europe      2007  26.3
...
```

We may want to measure the area fold change relative to 1980 for each region:

```
irrigation %>%
  group_by(region) %>%
  mutate(area_std_1980 = area/area[year == 1980])
# A tibble: 16 x 4
# Groups:   region [4]
   region      year  area area_std_1980
   <chr>      <dbl> <dbl>         <dbl>
1 Africa      1980   9.3             1
2 Africa      1990  11             1.18
3 Africa      2000  13.2           1.42
4 Africa      2007  13.6           1.46
5 Europe      1980  18.8             1
6 Europe      1990  25.3           1.35
7 Europe      2000  26.7           1.42
8 Europe      2007  26.3           1.40
...
```

Just like with `mutate()`, we can add more transformations, like the percentage change over each time point:

```
> irrigation <- irrigation %>%
+   group_by(region) %>%
+   mutate(area_std_1980 = area/area[year == 1980],
+          area_per_change = c(0, diff(area)/area[-length(area)] * 100))
> irrigation
# A tibble: 16 x 5
# Groups:   region [4]
   region      year  area area_std_1980 area_per_change
   <chr>      <dbl> <dbl>         <dbl>           <dbl>
1 Africa      1980   9.3             1               0
2 Africa      1990  11             1.18            18.3
3 Africa      2000  13.2           1.42            20.0
4 Africa      2007  13.6           1.46            3.03
5 Europe      1980  18.8             1               0
6 Europe      1990  25.3           1.35            34.6
7 Europe      2000  26.7           1.42            5.53
8 Europe      2007  26.3           1.40           -1.50
...
```

Reiterations Redo

Notice that we didn't need any looping in the preceding examples. You may have intuitively wanted to apply a for loop to calculate aggregation or transformation functions for each region, but it's not necessary. Avoiding for loops is somewhat of a past time in R, and is found in the base package with the apply family of functions.

Because vectorization is so fundamental to R, there's a bit of an unofficial contest to see how few for loops you can write. We imagine some useRs have a wall sign: "Days since last for loop:" like factories have for accidents.

This means there are some very old methods for reiterating tasks, along with some newer methods that make the process more convenient.

The old-school method relies on the apply family of functions, listed in Table 2-5. Except for apply(), pronounce them all as the first letter and then apply, hence "t apply" not "tapply."

Table 2-5. Base package apply family

Function	Use
apply()	Apply a function to each row or column of a matrix or data frame
lapply()	Apply a function to each element in a list
sapply()	Simplify the output of lapply()
mapply()	The multivariate version of sapply()
tapply()	Apply a function to values defined by an index
emapply()	Apply a function to values in an environment

There's a bit of a trend to disavow these workhorses of reiteration but you'll still see them a lot, so they're worth getting familiar with. Doing so will also help you to appreciate why the Tidyverse arose. As an example, let's return to the aggregation functions we applied to the PlantGrowth data frame above. In the apply family of functions, we could have used:

```
> tapply(PlantGrowth$weight, PlantGrowth$group, mean)
 ctrl trt1  trt2
5.032 4.661 5.526
> tapply(PlantGrowth$weight, PlantGrowth$group, sd)
     ctrl      trt1      trt2
0.5830914 0.7936757 0.4425733
```

You can imagine reading this as "take the weight column from the PlantGrowth dataset, split the values according to the label in the group column in the PlantGrowth dataset, apply the mean function to each group of values, and then return a named vector."

Can you see how tedious this is if you want to add more functions on there? Named vectors can be convenient, but they are not really a typical way that you want to store important data.

One attempt to simplify this process was implemented in plyr, the precursor to dplyr. plyr is pronounced *plier*, like the small multifunctional hand-held tool. We use it as such:

```
library(plyr)

ddply(PlantGrowth, "group", summarize,
      avg = mean(weight))
```

This is still sometimes used today, but it has mostly been superseded by a data frame-centric version of the package, hence the *d* in dplyr (say d-plier):

```
library(dplyr)
PlantGrowth %>%
  group_by(group) %>%
  summarize(avg = mean(weight))
```

But to be clear, we could have returned a data frame with other very old functions:

```
> aggregate(weight ~ group, PlantGrowth, mean)
  group weight
1  ctrl  5.032
2  trt1  4.661
3  trt2  5.526
```

Wow, what a great function, right? This thing is super old! You'll still see it around, and why not? Once you wrap your head around it, it's elegant and gets the job done, even though it still only applies one function. However, the ongoing push to use a unified Tidyverse framework, which is easier to read and arguably easier to learn, means the ancient arts are fading into the background.

These functions have existed since the early days of R and reflect, intuitively, what statisticians do *all the time*. The *split* data into chunks, defined by some property (rows, columns, categorical variables, objects), then they *apply* some kind of action (plotting, hypothesis testing, modeling, etc.), and then they *combine* the output together in some way (data frame, list, etc.). The process is sometimes called *split-apply-combine*. Realizing that this process kept repeating itself started to clarify for the community how to start thinking about data and, indeed, how to actually organize data. From this the idea of "tidy" data was born.[4]

As a last example of iterations, you're probably familiar with the python `map()` function. An analogous function can be found in the Tidyverse purrr package. This is

---

4 If you want to read more about this topic, check out Hadley Wickham's paper (*https://oreil.ly/LWXFa*).

convenient for reiterating over lists or elements in a vector, but it's beyond the scope of this book.

Final Thoughts

In Python, you often hear about *the* Python way ("Pythonic"). This means the proper Python syntax and the preferred method to perform a specific action. This doesn't really exist in R; there are many ways to go about the same thing, and people will use all varieties! Plus, they'll often mix dialects. Although some dialects are easier to read than others, this hybridization can make it harder to get into the language.

Added to this is the constant tweaking of an expanding Tidyverse. Functions are tagged as experimental, dormant, maturing, stable, questioning, superseded, and archived. Couple that with relatively lax standards for project-specific package management or for the use of virtual environments, and you can imagine a certain amount of growing frustration.

R officially celebrated its 20th birthday in 2020, and its roots are much older than that. Yet, it sometimes feels like R is currently experiencing a teenage growth spurt. It's trying to figure out how it suddenly got a lot bigger and can be both awkward and cool at the same time. Blending the different R dialects will take you a long way in discovering its full potential.

## CHAPTER 3

Python for UseRs

Rick J. Scavetta

Welcome, brave useR, to the wonderful world of the Pythonista! For many useRs, this brave new world may appear more varied—and thus more inconsistent and confusing—than what they're used to in R. But don't fret over diversity—celebrate it! In this chapter, I'll help you navigate through the rich and diverse Python jungle, highlighting various paths (workflows) that your Python-using colleagues may have taken and that you may choose to explore later on. Meanwhile, know that you'll eventually find the path that best suits you and your work environment; this will change over time and may not be the one outlined here. Like any good trek, use this route as a guide, not a rule book.

I'll cover the essentials of the four elements I mentioned in the introduction to this part: *functions*, *objects*, *logical expressions*, and *indexing*. But I'll begin by addressing three questions:

Question 1
   Which version and build (distribution) to use? There are a few different versions and builds of Python to choose from, in contrast to R.

Question 2
   Which tools to use? The wide variety of IDEs, text editors, and notebooks, plus the many ways of implementing virtual environments, adds more choices to make.

Question 3
   How does Python *the language* compare to R *the language*? Wrapping your head around an OOP-centric world, with a host of classes, methods, functions, and keywords provides another barrier to entry.

I'll address each of these questions in turn. My goal is to get you comfortable enough with reading and writing Python so you can continue your bilingual journey in Part III and Part IV. I'm not setting out to provide a full-fledged, deep introduction to Python for Data Science. For that purpose, see O'Reilly's *Python for Data Analysis* by Wes McKinney and *Python Data Science Handbook* by Jake VanderPlas; this chapter will help you appreciate those books even more.

If you're eager to get on with it and start using Python, you can skip to the section on notebooks, "Notebooks" on page 57, and visit the Google Colab Notebook (*https://oreil.ly/hLi6i*) for the lesson on Python, or access this chapter's script at our book repository on GitHub (*https://github.com/moderndatadesign/PyR4MDS*).

Versions and Builds

Although there are a few different distributions of R, useRs mostly stick with vanilla R obtained from r-project.org (*https://www.r-project.org*).[1] For Python, there are at least four common Python *builds* (aka distributions) to contend with. In each case you'll also want to consider the Python *version* as well.

First, you'll notice that you likely have a system version of Python already installed. On my machine, running macOS Big Sur (v11.1), I see this version of Python using the following terminal command:

```
$ python --version
Python 2.7.16
```

Interestingly, macOS also has python3 built in:

```
$ python3 --version
Python 3.8.6
```

These are the Python installations that macOS uses internally; there's no need to touch them.

Second, we have *vanilla* Python—the bare-bones, straight-from-the-source version of Python. At the time of writing, this is version 3.9. Version 2.x is no longer supported, and you should be using 3.x for future data science projects. Until you're sure all packages you'll use are compatible with the latest version, it's a safe bet to stick to the last minor update, 3.8 in this case. Indeed, you may have multiple minor versions on your system.

To install the specific version you want, visit the Python website (*https://www.python.org*) and follow the instructions on the download page.

---

1 Indeed, the other builds are seldom mentioned in proper company.

Installation varies depending on your system. As such, the official Python Setup and Usage guide (*https://oreil.ly/xZEhZ*) is the authoritative resource. If you encounter installation issues, a good starting point is to perform a literal web search (encase in double quotes) for the generic part of the error message.

Table 3-1 provides other sources, but you're well advised to just go to the source.[2]

Table 3-1. Installing Python

Platform	Site	Alternative
Linux	*python.org*	Python 3 is already installed.
macOS	*python.org*	Use `brew install python3` in the terminal.
Windows	*python.org*	Install Python from the Windows Store.

Third, there are two common Conda builds: Anaconda (aka Conda) and Miniconda. Conda offers package, dependency, and environment management for several programming languages, including Python and R, although it is seldomly used for R. These open source builds include Python, a suite of packages useful for data science, and a collection of IDEs (including RStudio). Anaconda comes in a free individual version plus various commercial versions. As the name suggests, Miniconda (*https://oreil.ly/oBi4M*) is a minimal installer. We'll see Miniconda make a reappearance in the last part of the book.

The Anaconda website (*https://oreil.ly/lQ6s6*) has detailed instructions for installation. You'll notice that Anaconda may not come packaged with the latest version of Python. For example, at the time of writing, Anaconda comes packaged with Python 3.8, not 3.9. So this provides some justification for our preference of vanilla Python 3.8, as mentioned previously. Anaconda is a popular build, but for our purposes we'll stick with vanilla Python to avoid the extra bells and whistles that, at this point, would only serve to distract us. Thus, I won't consider this option further but will mention some significant differences as needed if you choose to go down this path.

Fourth, you may decide to not use a local Python installation and instead use the popular online version of Python provided by Google Colab Notebooks (*https://oreil.ly/USncL*).[3] There are other online notebook tools, but it's beyond the scope of this book to detail all of them. Notebooks are akin to R Markdown documents but JSON-based. We'll discuss them in more detail later on.

I bet you can already guess that this early-stage diversity can result in downstream confusion when installation-specific issues arise. Moving forward, we'll assume you have a local or cloud-based installation of Python ready to go.

---

2 You'll need to install Homebrew (*https://brew.sh*) if you want to go that route for macOS.

3 You'll need a Google account to access this free resource.

Standard Tooling

Similar to R, there are many ways to access Python. Common methods include using the command line, IDEs, cloud-based IDEs, text editors, and notebooks. For simplicity, I'm not going to focus on executing Python on the command line. If you're familiar with executing scripts on the command line, this is familiar territory. If not, you'll cross that bridge when you come to it.

IDEs include JupyterLab, Spyder, PyCharm, and our beloved RStudio. Cloud-native IDEs include AWS Cloud9. These are all variations on a theme, and in my experience are not typically favored by Pythonistas, although there is a trend toward using cloud-based tools. It may sound strange that IDEs are not that popular. Why not use an IDE if you have a great one available? I think the answer is twofold. First, no IDE managed to position itself as the dominant, de facto, choice among Pythonistas like RStudio has among useRs. Second, because Python use cases are so varied, including often being executed on the command line itself, coding with IDEs just wasn't attractive for many Pythonistas, especially if they came from a coding background and felt comfortable without an IDE. For me, this feeds a bit into a narrative that says Python is both more difficult and better than R. Both are incorrect! Sorry :/ Nonetheless, you may be tempted to begin using Python with a comfortable-looking IDE. Here, we make the argument that text editors will serve you better in the long run. We'll get back to RStudio in the last part of the book as we bring Python and R together in a single script. For now, try to resist the urge to default to an IDE but watch for developments in cloud platforms that may direct future trends.

Text editors are the most common and seemly preferred tool for composing pure Python scripts. There are a host of fantastic text editors to choose from, waxing and waning in popularity year-on-year. Sublime (*https://www.sublimetext.com*), Atom (*https://atom.io*), Visual Studio Code (VS Code) (*https://code.visualstudio.com*), and even the classic editors Vim (*https://www.vim.org*) and Emacs (*https://oreil.ly/pGjVX*), among many others, are in common use. Nonetheless, VS Code, an open source editor developed and strongly supported by Microsoft, has emerged in the past few years as the top choice. A marketplace for extensions means that this editor provides strong and easy support for a variety of languages, including Python and R.[4] Thus, we'll focus on VS Code. Your first exercise is to obtain and install VS Code.

---

4 Again, although R is supported, useRs seldom work in VS Code.

The first time you open VS Code, you'll be greeted with the welcome screen where you can choose your theme (light, in our case), and install the Python language extension right away, as shown in Figure 3-1.

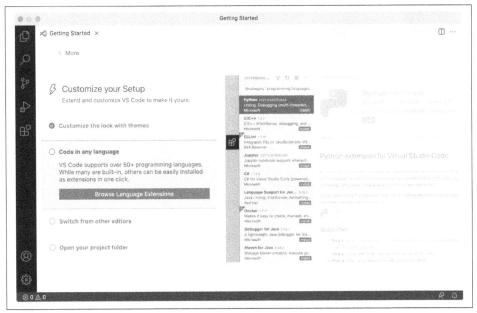

Figure 3-1. The VS Code welcome screen

When you click on the blank document icon in the upper left, you'll be requested to open a folder or clone a Git repository (from GitHub, for example), as shown in Figure 3-2. If you've cloned the book repo, use that; otherwise, use any folder you like. This is like opening a project in RStudio.

You'll also see files in the folder in the left sidebar. Open *ch03-py4r/py4r.py*, as in Figure 3-3.

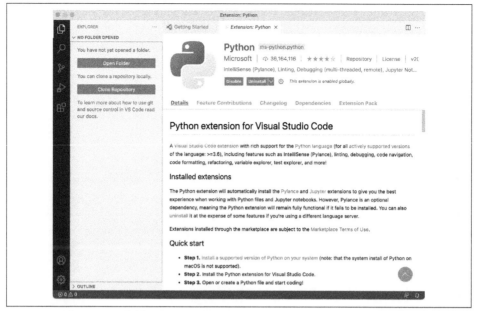

Figure 3-2. Opening a folder as a workspace

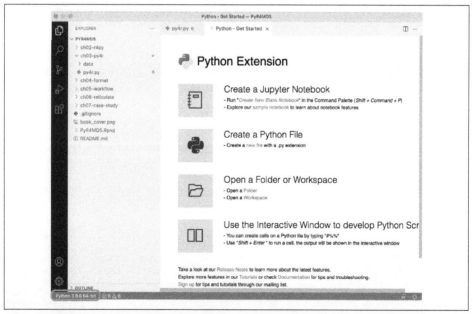

Figure 3-3. An opened folder, Python script, and extension guide

Because of the file extension, VS Code has automatically detected that you want to use a Python interpreter for this document. VS Code, like many other text editors, can execute code directly from the document if it knows which interpreter to use. You'll be greeted with the extension welcome page, and the blue footer notes the Python version that you're using. Remember that you may have many versions installed on your system. Here, I'm using v3.8.6.

The first item on the extension's welcome page is "Create a Jupyter Notebook." We'll get to that soon enough; for now, it's worth noting that we can use VS Code for both scripts and notebooks. Also note that the first bullet point in that item tells us that to open a notebook we should run a command in the Command Palette. You can access the Command Palette with the keyboard shortcut Shift-Command-P on a Mac (or Shift + Ctrl + P on a PC).[5] Return to our script and open the Command Palette. This is where you'll execute all variety of commands to make your life as a Pythonista easier. The Command Palette is a relatively new feature in RStudio but has been a standard way to navigate text editors for quite a while. Each extension you install from the marketplace will add more commands that you can access here. Our first command will be to "Create New Integrated Terminal (in Active Workspace)" (Figure 3-4). You can get this by beginning to type the command and then let autocomplete work its magic. Make sure you choose the "(in Active Workspace)" option. Remember, this is like an RStudio project, so we want to remain in our Active Workspace.

So by now we've settled on a text editor and we have our first (empty) Python script. It looks like we're ready to go—but not quite! We must address two crucial factors that you'll encounter each time you want to create a new Python project:

- Virtual (development) environment
- Installing packages

---

5 Astute userRs may have noticed that a Command Palette, invoked with the same keyboard shortcuts, was added to RStudio v1.4 in late 2020.

Figure 3-4. Accessing the Command Palette

Virtual Environments

Most useRs are accustomed to using RStudio projects, which keep the working directory tied to the project directory. These are convenient, in that we don't need to hard-code paths and are encouraged to keep all data and scripts in one directory. You'll already have that when opening an entire folder in a VS Code workspace.

A major downside of RStudio projects and VS Code workspaces is that they don't provide portable, reproducible development environments! Many useRs have a single, global installation of each package (see .libPaths()) and rarely specify the R version.

Now, dear useR, let's be honest with each other: if you haven't already, at some point you'll encounter the problem of package version conflicts. You've updated a package globally and now an old script is defunct because it's calling a deprecated function, or using a function's default arguments that have since changed, or for any number of other reasons due to package version conflicts. This is a surprisingly common occurrence in R and is a truly dismal practice when working over a long period of time or collaboratively. There have been many attempts to implement some kind of controlled development environment in R over the years. The most recent, and hopefully the solution that will finally stick, is renv. If you haven't kept up with developments there, please visit the package website from RStudio (*https://oreil.ly/xGVcr*).

Pythonistas have long used virtual environments to maintain the future compatibility of their projects, a sign of the programming-first approach of Python's origins. Here, a virtual environment is simply a hidden subdirectory within a project folder, called for example, *.venv*. The . is what makes it hidden. You have many hidden files and directories all over your computer, and for the most part they're hidden because you have no business sticking your fingers in there. Inside *.venv* you'll find the packages used in *this* specific project, and information about which Python build *this* project uses. Since each project now contains a virtual environment with all the packages and the appropriate package versions (!), you're guaranteed that the project will continue working indefinitely, so long as that virtual environment exists. We can visualize the potential dependency issues between different machines as in Figure 3-5, which highlights the benefit of having a single "source of truth" regarding package versions.

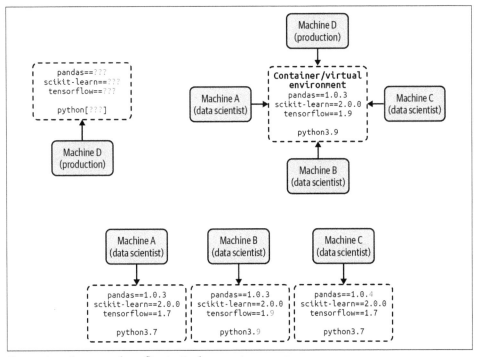

Figure 3-5. Sources of conflict in Python environments

Like everything in Python, there are many ways to make a virtual environment. We can use the venv or virtualenv packages. If you're using Anaconda, you'll use the Conda alternative, which we won't cover here. There are some subtle differences between venv and virtualenv, but at this point in the story they are irrelevant; let's just stick with venv. In your new terminal window, execute one of the commands in Table 3-2 depending on your platform, as I've done in Figure 3-6.

Table 3-2. Creating (and activating) a virtual environment with venv

Platform	Create	Activate (prefer VS Code autoactivate)
macOS X and Linux	`python3 -m venv .venv`	`source .venv/bin/activate`
Windows	`py -3 -m venv .venv`	`.venv\scripts\activate`

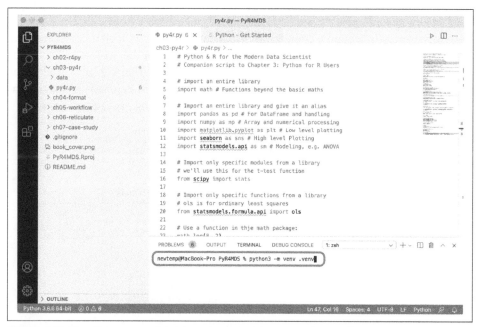

Figure 3-6. Creating a new virtual environment in our active workspace

After *creating* the virtual environment, you must *activate* it. The terminal commands for this are given in Table 3-2, but it's more convenient to let VS Code do what it does best. It will automatically detect your new virtual environment and ask if you want to activate it (see Figure 3-7). Go for it! Notice that the Python interpreter in the lower left will also explicitly mention (.venv): venv and in your open folder you'll have two hidden folders, *.venv* and *.vscode* (Figure 3-7).

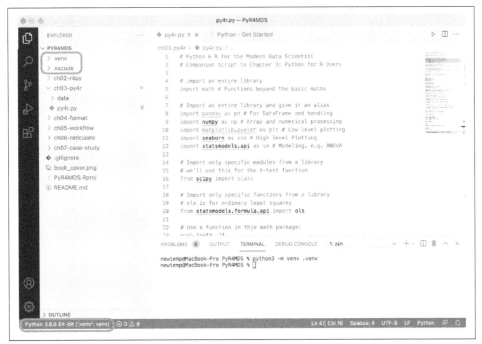

Figure 3-7. An active virtual environment

We'll get to package installation in a second; for now let's try to execute our first "Hello, world!" command. Type the following command in your script:

```
#%%
print('Hello, world!')
```

Actually the `print` is not necessary, but it makes what we're trying to do explicit. That looks a lot like a simple R function, right? The `#%%` is also not necessary, but it's a lovely feature of the Python extension in VS Code and is highly recommended! Typing `#%%` allows us to break up our long script into executable chunks. It's akin to R Markdown chunks, but much simpler and used in plain Python scripts. To execute the command, press Shift-Enter or click Run Cell, as shown in Figure 3-8.

You'll be promptly asked to install the `ipyKernel`. Do that, and you'll get the output present in the new upper-right pane, visible in Figure 3-9.

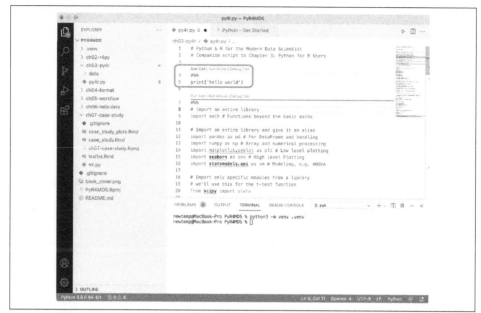

Figure 3-8. Executing your first code in Python

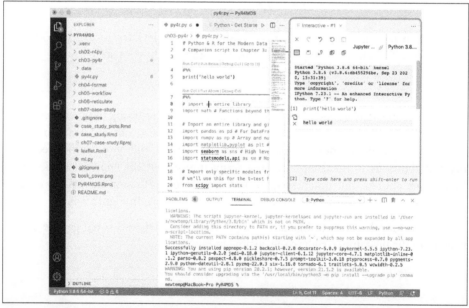

Figure 3-9. Viewing command output in the interactive Python viewer

Alright, *now* we're in business. That seems like a lot of work, but once you do it a couple times you'll develop a routine and get the hang of it!

Installing Packages

So far in the story we have installed some version of Python and accessed a workspace, like an R project, from within VS Code. We've also created a virtual environment, which we're now ready to populate with our favorite data science packages. If you took the Conda route, you'll have used different commands but you'll also be set to go with the most common data science packages preinstalled. That sounds really nice, but you may find that when you need to collaborate with other Python developers, e.g., data engineers or system administrators, they probably won't be using Anaconda. We feel that there's also something to be said for getting to the heart of Python without all the bells and whistles that Anaconda provides. Thus we've gone the vanilla route.

Before we get into packages, let's review some necessary terminology. In R, a library is a collection of individual packages. The same holds true for Python, but the use of *library* and *package* is not as strict. For example, pandas, the package that provides the `pd.DataFrame` class of objects, is referred to as both a library and a package on the pandas website itself. This mixture of the terms is common among Pythonistas, so if you're a stickler for names, don't let it confuse or bother you. However, *modules* are useful to note. A package is a collection of modules. This is useful to know since we can load an entire package or just a specific module therein. Thus, in general: library > package > module.

In R, you'd install packages from CRAN with the `install.packages()` function from within R itself. In Python, there are two equivalents of CRAN: PyPI (the Python Package Installer), for when using vanilla Python, and Conda, for when using Anaconda or Miniconda (we'll also see how to install packages in Google Colab directly in the online notebook later on). To install packages from PyPI using vanilla Python, you'll have to execute a command in the terminal. Recall that we still have our active terminal window in VS Code open from earlier. Execute the command `pip install matplotlib` in the terminal to install the matplotlib package in your virtual environment, as depicted in Figure 3-10. `pip` is the package installer for Python, which also comes in various versions.

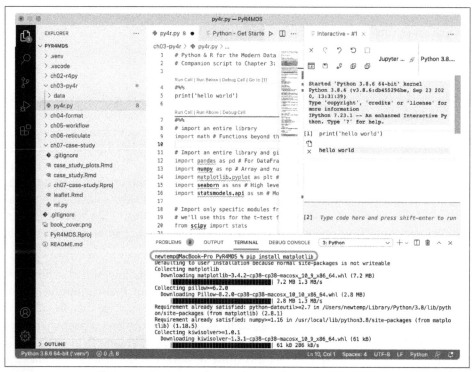

Figure 3-10. Installing a package into a virtual environment using the command line

Packages that you'll install in practically every virtual environment include NumPy, pandas, matplotlib, seaborn, and SciPy. You won't have to install all of them all the time since their package dependencies will take care of that. If they are already installed, `pip` will tell you and will not install anything further. The most common error messages you'll encounter here are when your package version is incompatible with your Python version. For this, you can either use a different Python kernel (the Python execution backend) for your project, or specify the exact package version you want to install. Like in R, you just need to install the package once, but you'll need to *import* it (i.e., *initialize* it) every time you activate your environment. It seems convenient that package installation is done in the terminal, separate from importing in the script. You've probably seen many stray `install.packages()` functions in R scripts, which is kind of annoying.

There are two more important points I want to mention. First, check all the packages installed in your environment, and their versions, in the terminal with:

```
$ pip freeze
```

Second, pipe this output to a file called *requirements.txt* by executing the following command:

```
$ pip freeze > requirements.txt
```

Other users can now use *requirements.txt* to install all the necessary packages by using the following command:

```
$ pip install -r requirements.txt
```

Notebooks

If you've followed the tutorial thus far, you're ready to proceed to the third question and begin exploring the Python language. Nonetheless, it's worthwhile reviewing notebooks, so read on. If you had a hard time getting Python set up locally, don't fret! Jupyter Notebooks is where you can take a deep breath, set your installation issues aside, and jump in afresh.

Jupyter Notebooks are built on the backbone of IPython, which originated in 2001. Jupyter, which stands for JUlia, PYThon, and R, now accommodates dozens of programming languages and can be used in the JupyterLab IDE or straight-up Notebooks in Jupyter. Notebooks allow you to write text using markdown, add code chunks, and see the inline output. It sounds a lot like R Markdown! Well, yes and no. Under the hood an R Markdown is a flat text file that gets rendered as an HTML, DOC, or PDF. Notebooks are exclusively JSON-based HTML and can natively handle interactive components. For useRs, this is kind of like an interactive R Markdown with a `shiny` runtime by default. This means that you don't compose a notebook as a flat text file, which is an important distinction when considering editing potential.

Coding in Python often consists of pure notebooks. For example, if you venture into cloud platforms that work with big data for machine learning (ML), like AWS Sage-Maker, Google AI Platform, or Azure Machine Learning Studio, you'll start with notebooks. As we've already seen, they're supported by VS Code. Other online versions include Kaggle competitions and published Jupyter Notebooks. Another variety of online notebooks is found in the Google Colab service (see Figure 3-11). This allows you to produce and distribute online notebooks with a Python backend and is what we'll use for exploring notebooks.

To get familiar working with notebooks, use this online tutorial from Jupyter (*https:// jupyter.org/try*). Just click on the Notebook Basics panel (*https://oreil.ly/QXfL6*) and pay particular attention to the keyboard shortcuts.

Open the Google Colab Notebook for this chapter (*https://oreil.ly/UVyP9*) to follow along.

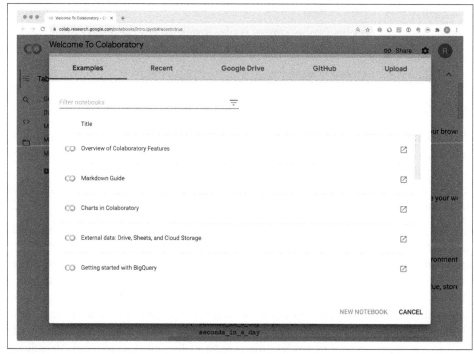

Figure 3-11. Examples for getting started with Python notebooks in using Google Colab

How Does Python, the Language, Compare to R?

By now you should have followed one of two paths. If you have installed Python locally, you should have:

1. A project directory where you'll store your data and script files.
2. A virtual environment set up within that directory.
3. The typical packages for data science installed in that virtual environment.

If you've decided to go the Google Colab route, you should have accessed this chapter's notebook.

Now it's time to start our project by importing the packages we'll use. Here, we'll see again that there are a variety of ways of doing this, but most are standard. Let's take a look. In the book's repository (*https://github.com/moderndatadesign/PyR4MDS*), you'll find a practice script with the following commands, or you can follow along in the Google Colab Notebook.

As we go through these commands, we'll introduce more new terminology—keywords, methods, and attributes—and discuss what they are in the context of Python.

First, we can import an entire package:

```
import math # Functions beyond the basic math
```

This allows us to use functions from the math package. The math package is already installed, so we didn't need to use pip, but we do need to import it.

This is the first time we encounter a common and important aspect of the Python language: *keywords*, which behave like R's reserved words but are more numerous. Right now there are 35 keywords in Python that can be placed in distinct groups (see the Appendix). Here import is an *import keyword*. As a useR accustomed to functional programming, you'd use library(math) in R. So, in this case, you can think of keywords as shortcuts to functions, which in many cases they are. That's just like operators in R (think <-, +, ==, &, etc.), which are just shortcuts to functions under the hood. They're not written in the classic function format, but they could be.

In short, keywords are reserved words that have very specific meanings. In this case, import stands in for a function to *import* all the functions from the math package. Many keywords act like this, but not all. We'll see some examples in a second.

But first, now that we have the functions from the math package, let's try this:

```
math.log(8, 2)
```

Here we see that the . has a specific meaning: inside the math package, access the log() function. The two arguments are the digit and base. So you can see why the R Tidyverse tends to use _ instead of . notation and why the prevalence of a meaningless . in many R functions frustrates many users coming from OOP-centric languages.

Second, we can import an entire package and give it a specific, typically standardized, alias:

```
import pandas as pd      # For data frame and handling
import numpy as np       # Array and numerical processing
import seaborn as sns    # High level plotting
```

There's our second keyword, as. Notice that it's not really acting as a stand in for a function unless we recall that <- is also a function. If we stretch our imaginations, we can imagine this is like the following in R:

```
dp <- library(dplyr)      # nonsense, but just as an idea
```

UseRs wouldn't ever do that, but it's the closest analogous command.[6] The as keyword is always used with import to provide a convenient alias for accessing a package

_____

6 But remember this for when we start using Python and R together because we'll see something very similar.

or module's functions.[7] Thus, it's an explicit way to call the exact function we want. Execute this function to import the dataset for future work:

```
plant_growth = pd.read_csv('ch03-py4r/data/plant_growth.csv')
```

Notice the . again? The preceding command is equivalent to this command in R:

```
plant_growth <- readr::read_csv("ch03-py4r/data/plant_growth.csv")
```

Third, we can import a specific *module* from a package:

```
from scipy import stats # e.g., for t-test function
```

There's our third keyword, from. It lets us go inside the SciPy package and *import* only the stats module.

Fourth, we can import a specific module from a package, also giving it a specific, typically standardized, alias:

```
import matplotlib.pyplot as plt # Low level plotting
import statsmodels.api as sm    # Modeling, e.g., ANOVA
```

Finally, we can also import only a specific function from a package:

```
from statsmodels.formula.api import ols # For ordinary least squares regression
```

Import a Dataset

We've seen how to import a dataset using a function from the pandas package:

```
plant_growth = pd.read_csv('ch03-py4r/data/plant_growth.csv')
```

Examine the Data

It's always good practice to look at our data before we start working on it. In R we'd use things like summary() and str(), or glimpse() if we had dplyr initialized. Let's see how that works in Python.

```
plant_growth.info()

<class 'pandas.core.frame.DataFrame'>
RangeIndex: 30 entries, 0 to 29
Data columns (total 2 columns):
 #   Column  Non-Null Count  Dtype
---  ------  --------------  -----
 0   weight  30 non-null     float64
 1   group   30 non-null     object
dtypes: float64(1), object(1)
memory usage: 608.0+ bytes
```

---

7 Referring back to log(), you're more likely to use np.log() instead of math.log() since it accepts a wider variety of input types.

```
plant_growth.describe()

        weight
count  30.000000
 mean   5.073000
  std   0.701192
  min   3.590000
  25%   4.550000
  50%   5.155000
  75%   5.530000
  max   6.310000

plant_growth.head()

   weight      group
0    4.17       ctrl
1    5.58       ctrl
2    5.18       ctrl
3    6.11       ctrl
4    4.50       ctrl
```

What the what?? This is the first time we've encountered this nomenclature, and there's that ever-present dot notation again! The functions `info()`, `describe()`, and `head()` are *methods* of the object `plant_growth`. A method is a function that is called by an object. As with other functions, we can also provide specific arguments, although in these cases we'll stick with the default.

Note in particular the output from the `info()` method. Here, we see for the first time that indexing begins at 0 in Python, as is the case with many programming languages —and why shouldn't it!? This is an important aspect of Pythonic programming. We'll see the consequences of this later on when we get to indexing.

The output from `.info()` also tells us that we have a `pd.DataFrame`. We'll explore different object classes soon.

How about looking at the shape (i.e., dimensions) and column names of the `plant_growth` object?

```
plant_growth.shape
```

```
(30, 2)
```

```
plant_growth.columns
```

```
Index(['weight', 'group'], dtype='object')
```

In this case, we are calling *attributes* of the object, so they don't receive any brackets. So here we see that any given object can call permissible methods and attributes, according to its class. You'll know this from R, when the class of an object allows it to be used in specific functions for which there are methods available for it. Under the hood, the same magic is happening. R is function-first, OOP-second. It's there, but we don't need to worry about it *as much* in functional programming. To give you an idea of how this works in R, consider a built-in dataset, sunspots. It's a ts class object (i.e., time series):

```
# in R
class(sunspots)
[1] "ts"
plot(sunspots)
# not shown
```

You can find the methods for the plot function using:

```
# in R
methods(plot)
```

There, you'll see the plot.ts() method, which is what is actually called when you provide a ts class object to the plot() function.

Finally, you may miss being able to actually *see* that dataset like we can with the RStudio view option. Not to worry! You can click on the table icon in the interactive Python kernel and see everything in your environment. If you click on the data frame, it will open up a view for you to examine it.

Data Structures and Descriptive Statistics

Alright, now that we've come to grips with methods and attributes, let's take a look at how you would generate some descriptive statistics. A pd.DataFrame is very similar to an R data.frame or tbl. It's a two-dimensional table where each column is a Series, like how columns are vectors of the same length in R data frames. Just like a pd.DataFrame itself, a Series also has methods and attributes. Recall that the group column is categorical. By now, this command should make sense to you:

```
plant_growth['group'].value_counts()

trt2    10
trt1    10
ctrl    10
Name: group, dtype: int64
```

The [ ] will be familiar to you from R; they index according to the name of a column. The . then takes this single column and calls a method, .value_counts(), which in this case counts the number of observations for each value.

How about this:

```
np.mean(plant_growth['weight'])
```

np says we're going to use a function from the NumPy package we imported earlier. Inside that function, we provide numerical values, the weight Series of the plant_growth data frame.

How about some summary statistics. Can you guess what this method will do?

```
# summary statistics
plant_growth.groupby(['group']).describe()
```

Just like with dplyr's `group_by()` function, the `groupby()` method will allow downstream methods to be applied on each subset according to a categorical variable, in this case the `group Series`. The `describe()` method will provide a suite of summary statistics for each subset.

This version is more specific:

```
plant_growth.groupby(['group']).agg({'weight':['mean','std']})
```

You can probably guess that the `.agg()` method stands for *aggregate*. Aggregation functions return a single value (typically), and in R we'd specify it using the `summarise()` function.

The input to the `.agg()` method, `{'weight':['mean','std']}`, is a dictionary (class `dict`). You can think of this as a key-value pair, defined here using `{}`:

```
{'weight':['mean','std']}
```

We could also have used the `dict()` function for the same purpose:

```
dict([('weight', ['mean','std'])])
```

Dictionaries are data storage objects in their own right, are part of standard vanilla Python, and as we see here are used as arguments to input in methods and functions. This is similar to how lists in R are used for both data storage and as a list of arguments in specific circumstances. Nonetheless, a dictionary is better thought of as an *associative array* since indexing is only by key, and not number. I may go so far as to say that a dictionary is even more like an environment in R, since that contains many objects but no indexing, but that may be a bit of a stretch.

Let's dig a bit deeper. The following commands produce the same output, but in different formats!

```
# Produces Pandas Series
plant_growth.groupby('group')['weight'].mean()

# Produces Pandas data frame
plant_growth.groupby('group')[['weight']].mean()
```

Notice the `[[]]` versus `[]`? It recalls a difference that you may have encountered in R when working with data frames that are not tibbles.

Data Structures: Back to the Basics

We've already seen three common types of data storage objects in Python, `pd.Data Frame`, `pandas Series`, and `dict`. Only `dict` is from vanilla Python, so before we move on, I want to look as some of the other basic structures: `lists`, `tuples`, and `NumPy arrays`. I'm introducing these much later than you'd expect; that's because I wanted to begin with data frames, which are intuitive and frequently used. So let's make sure we have the basics covered before we wrap up:

First, like in R, you'll see four key data types in Python:

Type	Name	Example
bool	Binary	True and False
int	Integer numbers	7,9,2,-4
float	Real numbers	3.14, 2.78, 6.45
str	String	All alphanumeric and special characters

Next, you'll encounter lists, one-dimensional objects. Unlike vectors in R, each element can be a different object, for example, another one-dimensional list. Here are two simple lists:

```
cities = ['Munich', 'Paris', 'Amsterdam']
dist = [584, 1054, 653]
```

Notice that the [ ] define a list. We actually already saw that when we defined the `dict` earlier:

```
{'weight':['mean','std']}
```

So both [] and {} alone are valid in Python and behave differently than in R. But remember, we did use [ ] earlier to index the data frame, which is very similar to R.

```
plant_growth['weight']
```

Finally, we have tuples, which are like lists, except they are immutable, i.e., unchangeable. They are defined by ( ), as such:

```
('Munich', 'Paris', 'Amsterdam')
```

A common use of tuples is when a function returns multiple values. As an example, the `divmod()` function returns the result of integer division and modulus of two numbers:

```
>>>   divmod(10, 3)
(3, 1)
```

The result is a tuple, but we can *unpack* the tuple and assign each output to a separate object:

```
int, mod = divmod(10, 3)
```

That's really convenient when defining custom functions. The equivalent in R would be to save the output to a list.

Astute useRs may be familiar with the multiple assign operator %<-%, introduced by the zeallot package and popularized by the Keras framework.

The last data structure I want to mention is the NumPy array. This is very similar to a one-dimensional list, but allows for vectorization, among other things. For example:

```
# A list of distances
>>> dist
[584, 1054, 653]
# Some transformation function
>>> dist * 2
[584, 1054, 653, 584, 1054, 653]
```

That's very different from what a useR would expect. If we were working on an NumPy array, we'd be back in familiar territory:

```
# Make a numpy array
>>> dist_array = np.array(dist)
>>> dist_array * 2
array([1168, 2108, 1306])
```

Indexing and Logical Expressions

Now that we have a variety of objects, let's look at how to index them. We already saw that we can use [] and even [[]] as we see in R, but there are a couple of interesting differences. Remember that indexing always begins at 0 in Python! Also, notice that one of the most common operators in R, :, makes a reappearance in Python but in a slightly different form. Here it's [start:end]:

```
>>> dist_array
array([ 584, 1054,  653])
>>> dist_array[:2]
array([ 584, 1054])
>>> dist_array[1:]
array([1054,  653])
```

The : operator doesn't need left- and right-hand sides. If one side is empty, the index begins or proceeds to the end. The start is *inclusive* and the end, if specified, is *exclusive*. Thus :2 takes index 0 and 1, and 1: takes index 1 up to the last element, which is unspecified and thus inclusive.

For two-dimensional data frames, we encounter the pandas .iloc, "index location" and .loc "location" methods:

```
# Rows: 0th and 2nd
>>> plant_growth.iloc[[0,2]]
   weight        group
0     4.17        ctrl
2     5.18        ctrl

# Rows: 0th to 5th, exclusive
```

```
# Cols: 1st
>>> plant_growth.iloc[:5, 0]
0    4.17
1    5.58
2    5.18
3    6.11
4    4.50
```

For the .loc(), we can introduce logical expressions, i.e., combinations of relational and logical operators to ask and combine True/False questions.

```
>>> plant_growth.loc[(plant_growth.weight <=  4)]
    weight        group
13    3.59        trt1
15    3.83        trt1
```

For more detail on indexing and logical expressions, see the notes in the Appendix.

Plotting

Alright, let's take a look at some data visualization of weight described by group. Since R's ggplot2 package will be recommended for elegant and flexible data visualization in Part III, I've left out the plots for the following commands. Nonetheless, it's useful to see the seaborn approach. Here we have a box plot:

```
sns.boxplot(x='group', y='weight', data=plant_growth)
plt.show()
# not shown
```

Just the points:

```
sns.catplot(x="group", y="weight", data=plant_growth)
plt.show()
```

And just the means with their standard deviations:

```
sns.catplot(x="group", y="weight", data=plant_growth, kind="point")
plt.show()
```

Notice that I'm using the seaborn package (alias sns) for data visualizations and then using the show() function from matplotlib to print the visualization to the screen.

Inferential Statistics

In this dataset, we have a specific set up in that we have three groups and we're interested in two specific two-group comparisons. We can accomplish this by establishing a linear model:

```
# fit a linear model
# specify model
model = ols("weight ~ group", plant_growth)

# fit model
results = model.fit()
```

We can get the coefficients of the model directly:

```
# extract coefficients
results.params.Intercept
results.params["group[T.trt1]"]
results.params["group[T.trt2]"]
```

Finally, let's take a look at a summary of our model:

```
# Explore model results
results.summary()
```

Alright, let's wrap this up by using a typical statistical test for this type of data: a one-way ANOVA. Notice that we're using our model, `results`, that we fitted previously:

```
# ANOVA
# compute anova
aov_table = sm.stats.anova_lm(results, typ=2)

# explore anova results
aov_table
```

If we want to do all pairwise comparisons, we can turn to Tukey's honestly significant differences (HSD) post hoc test:

```
from statsmodels.stats.multicomp import pairwise_tukeyhsd
print(pairwise_tukeyhsd(plant_growth['weight'], plant_growth['group']))
```

In this instance, we're starting with the *statsmodel* library, taking the stats package and the `multicomp` module therein, and extracting *from* that only the specific `pairwise_tukeyhsd()` function to *import*. In the second line, we execute the function with a continuous variable as the first argument and the grouping variable as the second argument.

Final Thoughts

In R, there has been a convergence on common practices and workflows since circa 2016. In Python, there is a lot more diversity in how to get up and running right from the word go. This diversity may seem daunting, but it's just a reflection of Python's origin story and use cases in the real world.

If you're a useR accustomed to the world of functional programming, wrapping your head around OOP methods can also seem pretty daunting, but once you get over that hurdle, you can start exploiting the power of Python where it truly shines, the topic of Part III.

Bilingualism II: The Modern Context

In this part you'll get your hands dirty and get a tour of applications of both languages in a modern context, in terms of the open source ecosystem and useful workflows.

These are the two dimensions we need to cover to get a coherent view. By going through both, you will obtain a clear picture when and where to use which language, open source package, and workflow.

Chapter 4

In this chapter, we'll go through how the variety of different data formats (i.e., image or text) are processed by different packages and which are the best ones.

Chapter 5

This chapter covers the most effective modern workflows (i.e., machine learning and visualization) for productive work for both R and Python.

Data Format Context

Boyan Angelov

In this chapter we'll review tools in Python and R for importing and processing data in a variety of formats. We'll cover a selection of packages, compare and contrast them, and highlight the properties that make them effective. At the end of this tour, you'll be able to select packages with confidence. Each section illustrates the tool's capabilities with a specific mini case study, based on tasks that a data scientist encounters daily. If you're transitioning your work from one language to another or simply want to find out how to get started quickly using complete, well-maintained, and context-specific packages, this chapter will guide you.

Before we dive in, remember that the open source ecosystem is constantly changing. New developments, such as transformer models (*https://oreil.ly/PLaGE*) and explainable artificial intelligence (XAI) (*https://oreil.ly/0sRHV*), seem to emerge every other week. These often aim at lowering the learning curve and increasing developer productivity. This explosion of diversity also applies to related packages, resulting in a nearly constant flow of new and (hopefully) better tools. If you have a very specific problem, there's probably a package already available for you, so you don't have to reinvent the wheel. Tool selection can be overwhelming, but at the same time this variety of options can improve the quality and speed of your data science work.

The package selection in this chapter can appear limited in view; hence, it is essential to clarify our selection criteria. So what qualities should we look for in a good tool?

It should be open source.

There is a large number of valuable commercial tools available, but we firmly believe that open source tools have a great advantage. They tend to be easier to extend and understand what their inner workings are, and are more popular.

It should be feature-complete.

The package should include a comprehensive set of functions that help the user do their fundamental work without resorting to other tools.

It should be well maintained.

One of the drawbacks of using open source software (OSS) is that sometimes packages have a short life cycle, and their maintenance is abandoned (so-called "abandonware"). We want to use packages that are actively worked on so we can feel confident they are up-to-date.

Let's begin with a definition. What is a *data format*? There are several answers available (*https://oreil.ly/M67LQ*). Possible candidates are data type, recording format, and file format. *Data type* is related to data stored in databases or types in programming languages (for example integer, float, or string). The *recording format* is how data is stored in a physical medium, such as a CD or DVD. And finally, what we're after, the *file format*, i.e., how information is *prepared for a computing purpose*.

With that definition in hand, you might still wonder why I dedicate an entire chapter to focus just on file formats. You have probably been exposed to them in another context, such as saving a PowerPoint slide deck with a *.ppt* or *.pptx* extension (and wondering which one is better). The problem here goes much further beyond basic tool compatibility. The way information is stored influences the complete downstream data science process. For example, if our end goal is to perform advanced analytics and the information is stored in a text format, we have to pay attention to factors such as character encoding (a notorious problem, especially for Python).[1] For such data to be effectively processed, it also needs to go through several steps, such as tokenization (*https://oreil.ly/ek4s9*) and stop word (*https://oreil.ly/zojPz*) removal.[2] Those same steps are not applicable to image data, even though we may have the same end goal in mind, e.g., classification. In that case other processing techniques are more suitable, such as resizing and scaling. These differences in data processing pipelines are shown in Figure 4-1. To summarize: the data format is the most significant factor influencing what you can and cannot do with it.

---

1 For a more thorough explanation on this, have a look at RealPython.com's guide (*https://oreil.ly/jQXLI*).

2 This is commonly referred to as *data lineage*.

We now use the word *pipeline* for the first time in this context, so let's use the opportunity to define it. You have probably heard the expression that "data is the new oil." This expression goes beyond a simple marketing strategy and represents a useful way to think about data. There are surprisingly many parallels between how oil and data are processed. You can imagine that the initial data that the business collects is the rawest form—probably of limited use initially. It then undergoes a sequence of steps, called data processing, before it's used in some application (i.e., for training an ML model or feeding a dashboard). In oil processing, this would be called refinement and enrichment—making the data usable for a business purpose. Pipelines transport the different oil types (raw, refined) through the system to its final state. The same term can be used in data science and engineering to describe the infrastructure and technology required to process and deliver data.

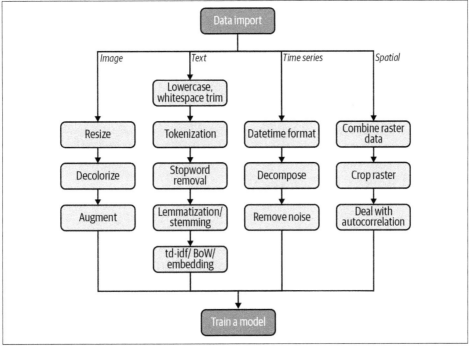

Figure 4-1. Difference between common data format pipelines (the lighter shade indicates the shared steps between the workflows)

Infrastructure and performance also need to be taken into consideration when working with a specific data format. For example, with image data, you'll need more storage availability. For time-series data you might need to use a particular database, such as InfluxDB (*https://oreil.ly/2crIr*). And finally, in terms of performance, image classification is often solved using deep learning methods based on convolutional neural networks (CNNs), which may require a graphics processing unit (GPU). Without it, model training can be very slow and become a bottleneck both for your development work and a potential production deployment.

Now that we have covered the reasons to carefully consider which packages to use, we'll have a look at the possible data formats. This overview is presented in Table 4-1 (note that those tools are mainly designed for small- to medium-size datasets). Admittedly, we are just scratching the surface on what's out there, and there are a few notable omissions (such as audio and video). Here, we'll focus on the most widely used formats.

Table 4-1. An overview of data formats and popular Python and R packages used to process data stored in them

Data type	Python package	R package
Tabular	pandas	readr, rio
Image	open-cv, scikit-image, PIL	magickr, imager, EBImage
Text	nltk, spaCy	tidytext, stringr
Time series	prophet, sktime	prophet, ts, zoo
Spatial	gdal, geopandas, pysal	rgdal, sp, sf, raster

This table is by no means exhaustive, and we are certain new tools will appear soon, but these are the workhorses fulfilling our selection criteria. Let's get them to work in the following sections, and see which ones are the best for the job!

External Versus Base Packages

In Chapter 2 and Chapter 3, we introduced packages very early in the learning process. In Python we used pandas right at the outset and transitioned to the Tidyverse in R relatively quickly. This allowed us to be productive much faster than if we went down the rabbit holes of archaic language features that you're unlikely to need as a beginner.[3] A programming language's utility is defined by the availability and quality of its third-party packages, as opposed to the core features of the language itself.

This is not to say that there aren't a lot of things that you can accomplish with just base R (as you'll see in some of the upcoming examples), but taking advantage of the open source ecosystem is a fundamental skill to increase your productivity and avoid reinventing the wheel.

---

3 Who else didn't learn what if __name__ == "__main__" does in Python?

Go Back and Learn the Basics

There is a danger in overusing third-party packages, and you have to be aware of when the right time to go back to the basics is. Otherwise you might fall victim to a false sense of security and become reliant on the training wheels provided by tools such as pandas. This might lead to difficulties when dealing with more specific real-world challenges.

Let's now see how this package versus base language concept plays out in practice by going into detail with a topic we're already familiar with: tabular data.[4] There are at least two ways to do this in Python. First, using pandas:

```
import pandas as pd

data = pd.read_csv("dataset.csv")
```

Second, with the built-in csv module:

```
import csv

with open("dataset.csv", "r") as file:  ❶
        reader = csv.reader(file, delimiter=",")
for row in reader:  ❷
        print(row)
```

❶ Note how you need to specify the file mode (*https://oreil.ly/KxuG2*), in this case "r" (standing for "read"). This is to make sure the file is not overwritten by accident, hinting at a more general-purpose oriented language.

❷ Using a loop to read a file might seem strange to a beginner, but it makes the process explicit.

This example tells us that pd.read_csv() in pandas provides a more concise, convenient, and intuitive way to import data. It is also less explicit than vanilla Python and not essential. pd.read_csv() is, in essence, a *convenience wrapper* of existing functionality—good for us!

Here we see that packages serve two functions. First, as we have come to expect, they provide *new* functionality. Second, they are also convenience wrappers for existing standard functions, which make our lives easier.

---

4 One table from the data, stored in a single file.

This is brilliantly demonstrated in R's rio package.[5] *rio* stands for "R input/output," and it does just what it says.[6] Here, the single function import() uses the file's filename extension to select the best function in a collection of packages for importing. This works in Excel, SPSS, stata, SAS, and many other common formats.

Another R Tidyverse package called vroom allows for fast import of tabular data and can read an entire directory of files in one command, with the use of map() functions or for loops.

Finally, the data.table package, which is often neglected at the expense of promoting the Tidyverse, provides the exceptional fread(), which can import very large files at a fraction of what base R or readr offer.

The usefulness of learning how to use third-party packages becomes more apparent when we try to perform more complex tasks, as we'll see next when processing other data formats.

Data Science from Scratch

Writing software from scratch is a great way to understand how things work under the hood. It's a recommended step in learning, especially after getting used to the tools at higher abstraction levels (such as scikit-learn). Excellent reading material on this topic is provided in Joel Grus's book *Data Science from Scratch* (O'Reilly, 2019).

Now that we can appreciate the advantages of packages, we'll demonstrate some of their capabilities. For this we'll work on several different real-world use cases, listed in Table 4-2. We won't focus on minute implementation details, but instead cover the elements that expose their benefits (and shortcomings) for the tasks at hand. Since the focus in this chapter is on data formats, and Chapter 5 is all about workflows, all these case studies are about data processing.

 For pedagogic purposes we have omitted parts of the code. If you'd like to follow along, executable code is available in the book's repository (*https://github.com/moderndatadesign/PyR4MDS*).

---

5 Not to forget readr, which was discussed in Chapter 2.

6 We did mention that statisticians are very literal, right?

Table 4-2. An overview of the different use cases

Data format	Use case
image	Swimming pool and car detection (*https://oreil.ly/0ajGP*)
text	Amazon product reviews processing (*https://oreil.ly/2n302*)
time series	Daily Australian temperatures (*https://oreil.ly/95slY*)
spatial	*Loxodonta africana* species distribution data (*https://www.gbif.org*)

Further information on how to download and process these data is available in the official repository (*https://github.com/moderndatadesign/PyR4MDS*) for the book.

Image Data

Images pose a unique set of challenges for data scientists. We'll demonstrate the optimal methodology by covering the challenge of aerial image processing—a domain of growing importance in agriculture, biodiversity conservation, urban planning, and climate change research. Our mini use case utilizes data from Kaggle collected to help the detection of swimming pools and cars. For more information on the dataset, you can use the URL in Table 4-2.

As we mentioned at the beginning of the chapter, downstream purpose influences data processing heavily. Since aerial data is often used to train machine learning algorithms, our focus will be on preparatory tasks.

The OpenCV (*https://opencv.org*) package is one of the most common ways to work with image data in Python. It contains all the necessary tools for image loading, manipulation, and storage. The "CV" in the name stands for *computer vision*—the field of machine learning that focuses on image data. Another handy tool that we'll use is scikit-image. As its name suggests, it's very much related to scikit-learn (*https://oreil.ly/ZCR55*).[7]

Here are the steps of our task (refer to Table 4-2):

1. Resize the image to a specific size.

2. Convert the image to black and white.

3. Augment the data by rotating the image.

For an ML algorithm to learn successfully from data, the input has to be cleaned (data munging), standardized (scaling), and filtered (feature engineering).[8] You can

---

7 This consistency is a common thread in the chapters in Part III and is addressed additionally in Chapter 5.

8 Remember—garbage in, garbage out.

imagine gathering a dataset of images (e.g., by scraping data from Google Images).[9] They will differ in some way or another—such as size and/or color. Steps 1 and 2 in our task list help us deal with that. Step 3 is handy for ML applications. The performance (i.e., classification accuracy, or area under the curve [AUC]) of ML algorithms depends mostly on the amount of training data, which is often in little supply. To get around this, without resorting to obtaining more data,[10] data scientists have discovered that playing around with the data already available, such as rotating and cropping, can introduce new data points. Those can then be used to train the model again and improve performance. This process is formally known as *data augmentation*.[11]

Enough talk—let's import the data! Check the complete code at the book's repository (*https://github.com/moderndatadesign/PyR4MDS*) if you want to follow along. The results are in Figure 4-2.

```
import cv2 ❶
single_image = cv2.imread("img_01.jpg")

plt.imshow(single_image)
plt.show()
```

❶ Using cv2 might seem confusing since the package is named OpenCV. cv2 is used as a shorthand name. The same naming pattern is used for scikit-image, where the import statement is shortened to skimage.

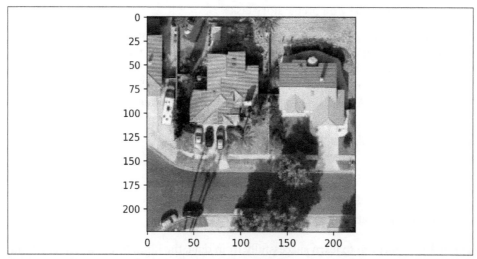

Figure 4-2. Raw image plot in Python with matplotlib

---

9 Using code to go through the content of a web page, download and store it in a machine-readable format.

10 Obtaining more data can be expensive, or even impossible in some cases.

11 If you want to learn more on data augmentation of images, have a look at this tutorial (*https://oreil.ly/YCQnP*).

So in what object type did `cv2` store the data? We can check with `type`:

```
print(type(single_image))
numpy.ndarray
```

Here we can observe an important feature that already provides advantages to using Python for CV tasks as opposed to R. The image is directly stored as a NumPy multi-dimensional array (nd stands for n-dimensions), making it accessible to a variety of other tools available in the wider Python ecosystem. Because this is built on the PyData stack, it's well supported. Is this true for R? Let's have a look at the magick package:

```
library(magick)
single_image <- image_read('img_01.jpg')
class(single_image)
[1] "magick-image"
```

The `magick-image` class is only accessible to functions from the magick package, or other closely related tools, but not the powerful base R methods (such as the ones shown in Chapter 2, with the notable exception of `plot()`). The differences in approaches of various open source packages supporting each other is illustrated in Figure 4-3, and is a common thread throughout the examples in this chapter.

 There is at least one exception to this rule—the EBImage package, a part of BioConductor (*https://bioconductor.org*). By using it you can get access to the image in its raw array form, and then use other tools on top of that. The drawback here is that it's part of a domain-specific package, and it might not be easy to see how it works in a standard CV pipeline.

Note that in the previous step (where we loaded the raw image in Python), we also used one of the most popular plotting tools, matplotlib (data visualization is covered in Chapter 5), so we again took advantage of this better design pattern.

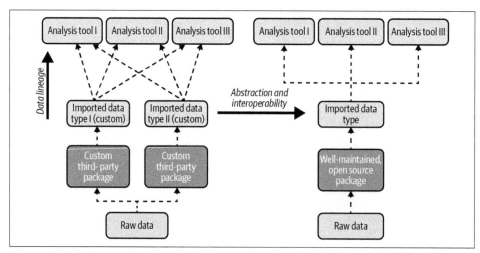

Figure 4-3. The two types of package design hierarchies as they are used during a data life cycle (bottom to top)

Now that we know that the image data is stored as a NumPy `ndarray`, we can use NumPy's methods. What's the size of the image? For this we can try the `.shape` method of `ndarray`:

```
print(single_image.shape)
224 224 3
```

It worked indeed! The first two output values correspond to the image `height` and `width` respectively, and the third one to the number of channels in the image—three in this case ((r)ed, (g)reen, and (b)lue). Now let's continue and deliver on the first standardization step—image resizing. Here we'll use `cv2` for the first time:

```
single_image = cv2.resize(single_image,(150, 150))
print(single_image.shape)
(150, 150, 3)
```

> If you gain experience working with such fundamental tools in both languages, you'll be able to test your ideas quickly, even without knowing whether those methods exist. If the tools you use are designed well (as in the better design in Figure 4-3), often they will work as expected!

Perfect, it worked like a charm! The next step is to convert the image to black and white. For this, we'll also use `cv2`:

```
gray_image = cv2.cvtColor(single_image, cv2.COLOR_RGB2GRAY)
print(gray_image.shape)
(150, 150)
```

The colors are greenish and not gray. This default option chooses a color scheme that makes the contrast more easily discernible for a human eye than black and white. When you look at the shape of the NumPy `ndarray`, you can see that the channel number has disappeared—there is just one now. Now let's complete our task, do a simple data augmentation step, and flip the image horizontally. Here we're again taking advantage that the data is stored as a NumPy array. We'll use a function directly from NumPy, without relying on the other CV libraries (OpenCV or scikit-image):

```
flipped_image = np.fliplr(gray_image)
```

The results are shown in Figure 4-4.

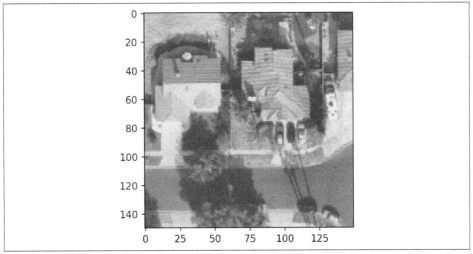

Figure 4-4. Plot of an image flipped by using NumPy functions

We can use scikit-image for further image manipulation tasks such as rotation, and even this different package will work as expected on our data format:

```
from skimage import transform
rotated_image = transform.rotate(single_image, angle=45)
```

The data standardization and augmentation steps we went through illustrate how the less complex package design (Figure 4-3) makes us more productive. We can drive the point home by showing a negative example for the third step, this time in R. For that, we'll have to rely on the adimpro package:

```
library(adimpro)
rotate.image(single_image, angle = 90, compress=NULL)
```

Whenever we load yet another package, we are decreasing the quality, readability, and reusability of our code. This issue is primarily due to possible unknown bugs, a steeper learning curve, or a potential lack of consistent and thorough documentation for that third-party package. A quick check on the status of adimpro on CRAN (*https://oreil.ly/Hr1f3*) reveals that the last time it was updated was in November 2019.[12] This is why using tools such as OpenCV, which work on image data by taking advantage of the PyData stack,[13] is preferred.

A less complex, modular, and abstract enough package design goes a long way to make data scientists productive and happy in using their tools. They are then free to focus on actual work and not dealing with complex documentation or a multitude of abandonware packages. These considerations make Python the clear winner in importing and processing image data, but is this the case for the other formats?

Text Data

The analysis of text data is often used interchangeably with the term *natural language processing* (NLP). This, in turn, is a subfield of ML. Hence it's not surprising to see that Python-based tools also dominate it. The inherently compute-intensive nature of working with text data is one good reason why that's the case. Another one is that dealing with larger datasets can be a more significant challenge in R than in Python (this topic is covered further in Chapter 5).[14] And it is a big data problem. The amount of text data has proliferated in recent years with the rise of services on the internet and social media giants such as Twitter and Facebook. Such organizations have also invested heavily in the technology and related open source tools, due to the fact that a large chunk of data available to them is in text format.

Similarly to the image data case, we'll start by designing a standard NLP task. It should contain the most fundamental elements of an NLP pipeline. For a dataset, we selected texts from the Amazon product reviews dataset (Table 4-2), and we have to prepare it for an advanced analytics use case, such as text classification, sentiment analysis, or topic modeling. The steps needed for completion are the following:

1. Tokenize the data.
2. Remove stop words.
3. Tag the parts of speech (PoS).

We'll also go through more advanced methods (such as word embeddings) in spaCy to demonstrate what the Python packages are capable of and, at the same time, provide a few R examples for comparison.

---

12 At the time of writing.

13 Not to be confused with the conference of the same name, the PyData stack refers to NumPy, SciPy, pandas, IPython, and matplotlib.

14 The R community has also rallied to the call and improved the tooling in recent times, but it still arguably lags behind its Python counterparts.

So what are the most common tools in Python? The most popular one is often referred to as the Swiss Army knife of NLP—the Natural Language Toolkit (NLTK).[15] It contains a good selection of tools covering the whole pipeline. It also has excellent documentation and a relatively low learning curve for its API.

As a data scientist, one of the first steps in a project is to look at the raw data. Here's one example review, along with its data type:

```
example_review = reviews["reviewText"].sample()
print(example_review)
print(type(example_review))
```

```
I just recently purchased her ''Paint The Sky With Stars''
 CD and was so impressed that I bought 3 of her previously
 released CD's and plan to buy all her music.  She is
 truely talented and her music is very unique with a
 combination of modern classical and pop with a hint of
 an Angelic tone. I still recommend you buy this CD. Anybody
  who has an appreciation for music will certainly enjoy her music.
```

```
str
```

This here is important—the data is stored in a fundamental data type in Python—str (string). Similar to the image data being stored as a multidimensional NumPy array, many other tools can have access to it. For example, suppose we were to use a tool that efficiently searches and replaces parts of a string, such as flashtext (*https://oreil.ly/JyYW6*). In that case, we'd be able to use it here without formatting issues and the need to coerce the data type.[16]

Now we can take the first step in our mini case study—*tokenization*. It will split the reviews into components, such as words or sentences:

```
sentences = nltk.sent_tokenize(example_review)
print(sentences)
```

```
["I just recently purchased her ''Paint The Sky With Stars''
CD and was so impressed that I bought 3 of her
previously released CD's and plan to buy all her music.",
 'She is truely talented and her music is very unique with
 a combination of modern classical and pop with a hint of an Angelic tone.',
 'I still recommend you buy this CD.',
 'Anybody who has an appreciation for music will certainly enjoy her music.']
```

Easy enough! For illustration purposes, would it be that hard to attempt this relatively simple task in R, with some functions from tidytext?

---

15 To learn more about NLTK, have a look at *Natural Language Processing with Python* by Steven Bird, Ewan Klein, and Edward Loper (O'Reilly), one of the most accessible books on working with text data.

16 Data type coercion is the conversion of one data type to another.

```
tidy_reviews <- amazon_reviews %>%
  unnest_tokens(word, reviewText) %>%
  mutate(word = lemmatize_strings(word, dictionary = lexicon::hash_lemmas))
```

This is one of the most well-documented methods to use. One issue with this is that it relies heavily on the "tidy data" concept, and also on the pipeline chaining concept from dplyr (we covered both in Chapter 2). These concepts are specific to R, and to use tidytext successfully, you would have to learn them first, instead of directly jumping to processing your data. The second issue is the output of this procedure—a new data.frame containing the data in a processed column. Although this might be what we need in the end, this skips a few intermediate steps and is several layers of abstraction higher than what we did with nltk. Lowering this abstraction and working in a more modular fashion (such as processing a single text field first) adheres to software development best practices, such as DRY ("Do not repeat yourself") and separation of concerns.

The second step of our small NLP data processing pipeline is removing stop words:[17]

```
tidy_reviews <- tidy_reviews %>%
  anti_join(stop_words)
```

This code suffers from the same issues, along with a new confusing function—anti_join. Let's compare it to the simple list comprehension (more information on this in Chapter 3) step in nltk:

```
english_stop_words = set(stopwords.words("english"))
cleaned_words = [word for word in words if word not in english_stop_words]
```

english_stop_words is just a list, and then the only thing we do is loop through every word in another list (words) and remove it *if* it's present in both. This is easier to understand. There's no relying on advanced concepts or functions that are not directly related. This code is also at the right level of abstraction. Small code chunks can be used more flexibly as parts of a larger text processing pipeline function. A similar meta processing function in R can become bloated—slow to execute and hard to read.

While nltk allows for such fundamental tasks, we'll now have a look at a more advanced package—spaCy. We'll use this for the third and final step in our case study —part of speech (PoS) tagging:[18]

```
import spacy

nlp = spacy.load("en_core_web_sm") ❶
```

---

17 This is a common step in NLP. Some examples of stop words are "the," "a," and "this." These need to be removed since they rarely offer useful information for ML algorithms.

18 The process of labeling words with the PoS they belong to.

```
doc = nlp(example_review) ❷
print(type(doc))

spacy.tokens.doc.Doc
```

❶ Here we are loading all the advanced functionality we need through one function.

❷ We take one example review and feed it to a spaCy model, resulting in the `spacy.tokens.doc.Doc` type, not a `str`. This object can then be used for all kinds of other operations:

```
for token in doc:
    print(token.text, token.pos_)
```

The data is already tokenized on loading. Not only that—all the PoS tags are marked already!

The data processing steps that we covered are relatively basic. How about some newer and more advanced NLP methods? We can take word embeddings, for example. This is one of the more advanced text vectorization methods,[19] where each vector represents the meaning of a word based on its context. For that, we can already use the same `nlp` object from the spaCy code:

```
for token in doc:
    print(token.text, token.has_vector, token.vector_norm, token.is_oov)

for token in doc:...
I True 21.885008 True
just True 22.404057 True
recently True 23.668447 True
purchased True 23.86188 True
her True 21.763712 True
' True 18.825636 True
```

It's a welcome surprise to see that those abilities are already built-in into one of the most popular Python NLP packages. On this level of NLP methods, we can see that there's almost no alternative available in R (or even other languages, for that matter). Many analogous solutions in R rely on wrapper code around a Python backend (which can defeat the purpose of using the R language).[20] This pattern is often seen in the book, especially in Chapter 5. The same is also true for some other advanced methods such as transformer models.[21]

---

19 Converting text into numbers for ingestion by an ML algorithm.

20 Such as trying to create custom embeddings. Check the RStudio blog (*https://oreil.ly/waD3o*) for more information.

21 You can read more about that on the RStudio blog (*https://oreil.ly/rUaWz*).

For Round 2, Python is again the winner. The capabilities of nltk, spaCy, and other associated packages make it an excellent choice for NLP work!

Time Series Data

The time-series format is used to store any data with an associated temporal dimension. It could be as simple as shampoo sales from a local grocery store, with a timestamp, or millions of data points from a sensor network measuring humidity in an agricultural field.

 There are some exceptions to the domination of R for the analysis of time-series data. The recent developments in deep learning methods, in particular, long short-term memory (LSTM) networks have proven to be very successful for time-series prediction. As is the case for other deep learning methods (more on this in Chapter 5), this is an area better supported by Python tools.

Base R

There are quite a few different packages that useRs can employ to analyze time-series data, including xts and zoo, but we'll be focusing on base R functions as a start. After this, we'll have a look at one more modern package to illustrate more advanced functionality: Facebook's Prophet (*https://oreil.ly/WNOyF*).

Weather data is both widely available and relatively easy to interpret, so for our case study, we'll analyze the daily minimum temperature in Australia (Table 4-2). To do a time-series analysis, we need to go through the following steps:

1. Load the data into an appropriate format.
2. Plot the data.
3. Remove noise and seasonal effects and extract trend.

Then we would be able to proceed with more advanced analysis. Imagine we have loaded the data from a *.csv* file into a data.frame object in R. Nothing out of the ordinary here. Still, differently from most Python packages, R requires data coercion into a specific object type. In this case, we need to transform the data.frame into a ts (which stands for time series).

```
df_ts <- ts(ts_data_raw$Temp, start=c(1981, 01, 01),
            end=c(1990, 12, 31), frequency=365)
class(df_ts)
```

So why would we prefer that to pandas? Well, even after you manage to convert the raw data into a time series pd.DataFrame, you'll encounter a new concept—data frame indexing (see Figure 4-5). To be efficient in data munging, you'll need to understand how this works first!

Index	A	B
2012-01-01 00:00:00	-6.088060	1.001294
2012-01-01 00:03:00	10.243678	1.074597
2012-01-01 00:06:00	-10.590584	0.987309
2012-01-01 00:09:00	11.362228	0.944953
2012-01-01 00:12:00	33.541257	1.095025
2012-01-01 00:15:00	-8.595393	1.035312

Figure 4-5. The time series index in pandas

This indexing concept can be confusing, so let's now look at what the alternative is in R and whether that's better. With the `df_ts` time series object, there are already a few useful things we can do. It's also a good starting point when you are working with more advanced time series packages in R because the coercion of a `ts` object into xts or zoo should throw no errors (this once again is an example of the good object design we covered in Figure 4-3). The first thing you can try to do is `plot` the object, which often yields good results in R:

```
plot(df_ts)
```

Calling the `plot` function does not simply use a standard function that can plot all kinds of different objects in R (this is what you would expect). It calls a particular method that is associated with the data object (more on the difference between functions and methods is available in Chapter 2). A lot of complexity is hidden behind this simple function call!

The results from `plot(df_ts)` in Figure 4-6 are already useful. The dates on the x-axis are recognized, and a `line` plot is chosen instead of the default `points` plot.

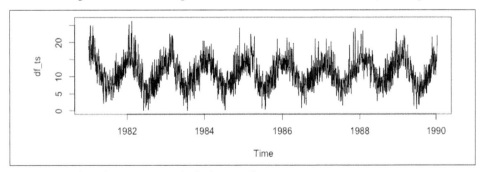

Figure 4-6. Plot of a time-series (ts) object in base R

The most prevalent issue in analyzing time-series data (and most ML data, for that matter) is dealing with noise. The difference between this data format and others is that there are a few different noise sources, and different patterns can be cleaned. This is achieved by a technique called *decomposition*, for which we have the built-in and well-named function decompose:

```
decomposed_ts <- decompose(df_ts)
plot(decomposed_ts)
```

The results can be seen in Figure 4-7.

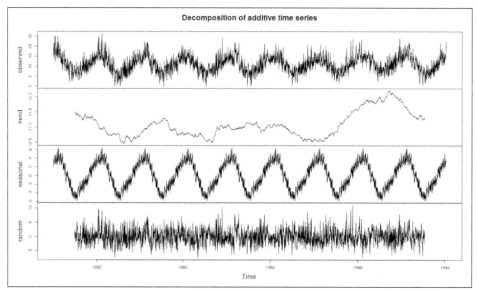

Figure 4-7. Plot of decomposed time-series in base R

We can see what the random noise is and also what is a seasonal and overall pattern. We achieved all this with just one function call in base R! In Python, we would need to use the statsmodels package to achieve the same.

Prophet

For analyzing time-series data, we also have another exciting package example. It's simultaneously developed for both R and Python (similar to the lime explainable ML tool): Facebook Prophet (*https://oreil.ly/WNOyF*). This example can help us compare the differences in API design. Prophet is a package whose main strength lies in the flexibility for a domain user to adjust to their particular need, ease of use of the API, and focus on production readiness. These factors make it a good choice for prototyping time series work and using it in a data product. Let's have a look; our data is stored as a pd.DataFrame in df:

```
from fbprophet import Prophet

m = Prophet()
m.fit(df) ❶

future = m.make_future_dataframe(periods=365) ❷
future.tail()
```

❶ Here we see the same `fit` API pattern again, borrowed from scikit-learn.

❷ This step creates a new empty `pd.DataFrame` that stores our predictions later.

```
library(prophet)

m <- prophet(df)

future <- make_future_dataframe(m, periods = 365)
tail(future)
```

Both are simple enough and contain the same number of steps—this is an excellent example of a consistent API design (more on this in Chapter 5).

 It's an interesting and helpful idea to offer a consistent user experience across languages, but we do not predict it'll be widely implemented. Few organizations possess the resources to do such work, which can be limiting since compromises have to be made in software design choices.

At this point, you can appreciate that knowing both languages would give you a significant advantage in daily work. If you were exposed only to the Python package ecosystem, you would probably try to find similar tools for analyzing time series and miss out on the incredible opportunities that base R and related R packages provide.

Spatial Data

The analysis of spatial data is one of the most promising areas in modern machine learning and has a rich history. New tools have been developed in recent years, but R has had the upper hand for a long time, despite some recent Python advances. As in the previous sections, we'll look at a practical example to see the packages in action.

 There are several formats of spatial data available. In this subsection, we are focusing on the analysis of *raster* data. For other formats there are some interesting tools available in Python, such as GeoPandas (*https://geopandas.org*), but this is out of scope for this chapter.

Our task is to process occurrence (location-tagged observations of the animal in the wild) and environmental data for *Loxodonta africana* (African elephant) to make it suitable for spatial predictions. Such data processing is typical in species distribution modeling (SDM), where the predictions are used to construct habitat suitability maps used for conservation. This case study is more advanced than the previous ones, and a lot of the steps hide some complexity where the packages are doing the heavy lifting. The steps are as follows:

1. Obtain environmental raster data.

2. Cut the raster to fit the area of interest.

3. Deal with spatial autocorrelation with sampling methods.

To solve this problem as a first step, we need to process raster data.[22] This is, in a way, very similar to standard image data, but still different in processing steps. For this R has the excellent raster package available (the alternative is Python's gdal and R's rgdal, which in our opinion, are trickier to use).

```
library(raster)
climate_variables <- getData(name = "worldclim", var = "bio", res = 10)
```

raster allows us to download most of the common useful spatial environmental datasets, including the bioclimactic data:[23]

```
e <- extent(xmin, xmax, ymin, ymax)
coords_absence <- dismo::randomPoints(climate_variables, 10000, ext = e)
points_absence <- sp::SpatialPoints(coords_absence,
                                    proj4string = climate_variables@crs)
env_absence <- raster::extract(climate_variables, points_absence)
```

Here we use the handy extent function to crop (cut) the raster data—we are only interested in a subsection of all those environmental layers surrounding the occurrence data. Here we use the longitude and latitude coordinates to draw this rectangle. As a next step, to have a classification problem, we are randomly sampling data points from the raster data (those are called pseudo absences). You could imagine that those are the 0s in our classification task, and the occurrences (observations) are the 1s—the target variable. We then convert the pseudo-absences to SpatialPoints, and finally extract the climate data for them as well. In the SpatialPoints function, you can also see how we specify the geographic projection system, one of the fundamental concepts when analyzing spatial data.

---

22 Data representing cells, where the cell value represents some information.

23 Environmental features that have been determined by ecologists to be highly predictive of species distributions, i.e., humidity and temperature.

One of the most common issues when working in ML is correlations within the data. The fundamental assumption for a correct dataset is that the individual observations in the data are *independent* of each other to get accurate statistical results. This issue is always present in spatial data due to its very nature. This issue is called *spatial auto-correlation*. There are several packages available for sampling from the data to mitigate this risk. One such package is ENMeval:

```
library(ENMeval)
check1 <- get.checkerboard1(occs, envs, bg, aggregation.factor=5)
```

The `get.checkerboard1` function samples the data in an evenly distributed manner, similar to taking equal points from each square from a black-and-white chessboard. We can then take this resampled data and successfully train an ML model without worrying about spatial autocorrelation. As a final step, we can take those predictions and create the habitat suitability map, shown in Figure 4-8.

```
raster_prediction <- predict(predictors, model)
plot(raster_prediction)
```

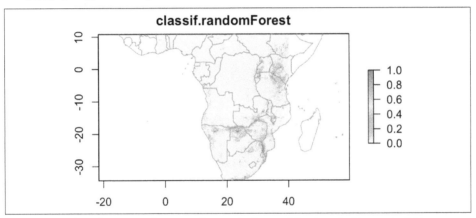

Figure 4-8. Plot of a raster object prediction in R, resulting in a habitat suitability map

When you're working with spatial raster data, the better package design is provided by R. The fundamental tools such as raster provide a consistent foundation for more advanced application-specific ones such as ENMeval and dismo, without the need to worry about complex transformation or error-prone type coercion.

Final Thoughts

In this chapter we went through the different common data formats and the best packages to process them so they are ready for advanced tasks. In each case study, we demonstrated a good package design and how that can make a data scientist more productive. We have seen that for more ML-focused tasks, such as CV and NLP, Python is providing the better user experience and lower learning curve. In contrast, for more time series prediction and spatial analysis, R has the upper hand. Those selection choices are shown in Figure 4-9.

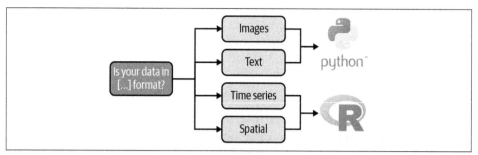

Figure 4-9. Decision tree for package selection

What the best tools have in common is the better package design (Figure 4-3). You should always use the optimal tool for the job and pay attention to the complexity, documentation, and performance of the tools you use!

## CHAPTER 5

Workflow Context

Boyan Angelov

A common source of frustration for data scientists is discussing their work with colleagues from adjacent fields. Let's take the example of someone who has been working primarily in developing ML models, having a chat about their work with a colleague from the business intelligence (BI) team, which is more focused on reporting. More often than not, such a discussion can make both parties uncomfortable due to a perceived lack of knowledge about each other's work domain (and associated workflows)—despite sharing the same job title. The ML person might wonder what D3.js is, the grammar of graphics, and all that. On the other hand, the BI data scientist might feel insecure about not knowing how to build a deployable API. The feelings that might arise from such a situation have been termed *impostor syndrome*, where doubts about your competency arise. Such a situation is a by-product of the sheer volume of possible applications of data science. A single person is rarely familiar to the same extent with more than several subfields. Flexibility is still often required in this fast-evolving field.

This complexity sets the foundation for the workflow focus in this chapter. We'll cover the primary data science workflows and how the languages' different ecosystems support them. Much like Chapter 4, at the end of this chapter, you'll have everything needed for making educated decisions regarding your workflows.

Defining Workflows

Let's take a step back and define a workflow:

> A *workflow* is a complete collection of tools and frameworks to perform all tasks required from a specific job function.

For this example, let's say you're an ML engineer. Your daily tasks might include tools to obtain data, process it, train a model on it, and deploying frameworks. Those, collectively, represent the ML engineer workflow. An overview of the data workflows for this and other titles and their supporting tools is presented in Table 5-1.

Table 5-1. Common data science workflows and their enabling tools.

Method	Python package	R package
Data munging[a]	pandas	dplyr
EDA	matplotlib, seaborn, pandas	ggplot2, base R, Leaflet
Machine learning	scikit-learn	mlr, tidymodels, caret
Deep learning	Keras, TensorFlow, PyTorch	Keras, TensorFlow, torch
Data engineering[b]	Flask, BentoML, FastAPI	plumber
Reporting	Jupyter, Streamlit	R Markdown, Shiny

[a] Data munging (or wrangling) is such a fundamental topic in data science that it was already covered in Chapter 2.
[b] There is much more to data engineering than model deployment, but we decided to focus on this subset to illustrate Python's ability.

We omitted some areas in the hope that the listed ones are the most common and critical. Those selected workflows are related to each other, as presented in Figure 5-1. This diagram borrows heavily from the CRISP-DM framework (*https://oreil.ly/19361*), which shows all significant steps in a typical data science project. Each of the diagram's steps has a separate workflow associated with it, generally assigned to an individual or a team.

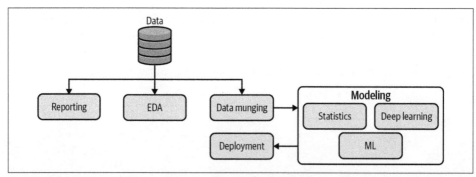

Figure 5-1. Metaworkflow in data science and engineering

Now that we have defined a workflow, what are the defining properties of a "good" one? We can compile a checklist with three main factors to consider:

1. It's well established. It's widely adopted by the community (also across different application domains, such as computer vision or natural language processing).

2. It's supported by a well-maintained, open source ecosystem and community. A workflow that relies heavily on closed-source and commercial applications (such as MATLAB) is not considered acceptable.

3. It's suitable for overlapping job functions. The best workflows are similar to Lego bricks—their modular design and extensibility can support diverse tech stacks.

With the big picture and definitions out of the way, let's dive deeper into the different workflows and how they are supported by R and Python!

Exploratory Data Analysis

Looking at numbers is *hard*. Looking at rows of data containing millions upon millions of them is even more challenging. Any person dealing with data faces this challenge daily. This need has led to considerable developments in data visualization (DV) tools. A recent trend in the area is the explosion of self-serving analytics tools, such as Tableau (*https://www.tableau.com*), Alteryx (*https://www.alteryx.com*), and Microsoft Power BI (*https://oreil.ly/5FMc2*). These are very useful, but the open source world has many alternatives available, often rivaling or even exceeding their commercial counterparts' capabilities (except, in some cases, ease of use). Such tools collectively represent the EDA workflow.

When to Use a GUI for EDA

Many data scientists frown at the notion of using a GUI for their daily work. They would much rather prefer the flexibility and utility of command-line tools instead. Nevertheless, one area where using a GUI makes more sense (for productivity reasons) is EDA. It can be quite time-consuming to generate multiple plots, especially at the beginning of a data science project. Usually, one would need to create tens, if not hundreds of them. Imagine writing the code for each one (even if you improve your code's organization by refactoring into functions). For some larger datasets, it's sometimes much easier to use some GUI, such as AWS QuickSight or Google Data Studio. By using a GUI, the data scientist can quickly generate a lot of plots first and only then write the code for the ones that make the cut after screening. There are a few good open source GUI tools, for example Orange (*https://orange.biolab.si*).

EDA is a fundamental step at the beginning of the analysis of any data source. It is typically performed directly after data loading, at the stage where there's a significant need for business understanding. This explains why it's an essential step. You are probably familiar with the *garbage in*, *garbage out* paradigm—the quality of any data project depends on the quality of the input data and the domain knowledge behind it. EDA enables the success of the downstream workflows (such as ML), ensuring both the data and the assumptions behind it are correct and of sufficient quality.

In EDA, R has far better tools available than Python. As we discussed in Chapter 1 and Chapter 2, R is a language made *by* statisticians and *for* statisticians (remember FUBU from Chapter 2?), and data visualization (plotting) has been of great importance in statistics for decades. Python has made some forward strides in recent years but is still seen as lagging (you need just to look at example matplotlib plot to realize this fact package also enables the creation of beautiful plots quickly but still lags behind the ggplot features).[1] Enough praise for R; let's have a look at why it's great for EDA!

Static Visualizations

You should already be acquainted with base R's powers in terms of DV from Chapter 4, especially regarding time series plotting. Here we'll take a step further and discuss one of the most famous R packages—ggplot2. It's one of the main reasons why Pythonistas want to switch to R.[2] What makes ggplot2 so successful in EDA work is that it's based on a well thought-through methodology—the grammar of graphics, which was developed by L. Wilkinson. The ggplot2 package was developed by Hadley Wickham.[3]

What *is* the grammar of graphics? The original paper (*https://oreil.ly/BFQQc*) behind it is titled "A Layered Grammar of Graphics," and the word "layered" holds the key. Everything you see on a plot contributes to a larger stack or system. For example, the axes and grids form a separate layer compared to the lines, bars, and points. Those latter elements constitute the data layer. The complete stack of layers forms the result—a complete `ggplot`. Such a modular design pattern allows for great flexibility

---

1 It's a bit unfair to present matplotlib as the only viable alternative from Python. The seaborn (*https://seaborn.pydata.org*) package also enables the creation of beautiful plots quickly but still lags behind the ggplot features. It's worth mentioning that newer versions of pandas have plotting capabilities as well, so we should watch this space.

2 There have been attempts to re-create this package in Python, such as ggplot (*https://oreil.ly/z4AoA*), but they have not caught on in the community so far.

3 He wrote many other packages, and in some ways almost single-handedly changed the way people use R in a modern context. Have a look at Chapter 2 and *The Grammar of Graphics* by Leland Wilkinson et. al. (Springer) for more information on his packages.

and provides a new way of thinking about data visualization. The logic behind grammar of graphics is illustrated in Figure 5-2.

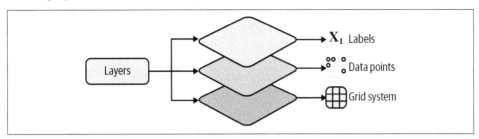

Figure 5-2. The layered grammar of graphics

To illustrate the different procedures for a regular EDA workflow, we'll use the star wars dataset (available from the dplyr package).[4] This dataset contains information on characters in the Star Wars movies, such as their gender, height, and species. Let's have a look!

```
library(ggplot2)
library(dplyr)

data("starwars") ❶
```

❶ This will make the dataset visible in your RStudio environment, but it's not strictly necessary.

As a first step, let's do a basic plot:

```
ggplot(starwars, aes(hair_color)) +
    geom_bar()
```

This plots the counts of the hair color variable. Here, we see a familiar operator, +, used unconventionally. We use + in ggplot2 to *add* layers on top of each other in ggplot2. Let's build on this with a more involved case. Note that we omitted a filtering step from the code here (there's an outlier—Jabba the Hutt): [5]

```
ggplot(starwars, aes(x = height, y = mass, fill = gender)) + ❶
    geom_point(shape = 21, size = 5) + ❷
    theme_light() + ❸
    geom_smooth(method = "lm") + ❹
    labs(x = "Height (cm)", y = "Weight (cm)",
        title = "StarWars profiles ",
        subtitle = "Mass vs Height Comparison",
        caption = "Source: The Star Wars API") ❺
```

---

4 More information on the dataset is available in the R Package Documentation (*https://oreil.ly/JXsJg*).

5 Did you know that his real name is Jabba Desilijic Tiure?

❶ Specify which data and features to use.

❷ Select a points plot (the most suitable for continuous data).

❸ Use a built-in theme—a collection of specific layer styles.

❹ Fit a linear model and show the results as a layer on the plot.

❺ Add title and axes labels.

The results of this plotting operation are shown in Figure 5-3. With just several lines of code, we created a beautiful plot, which can be extended even further.

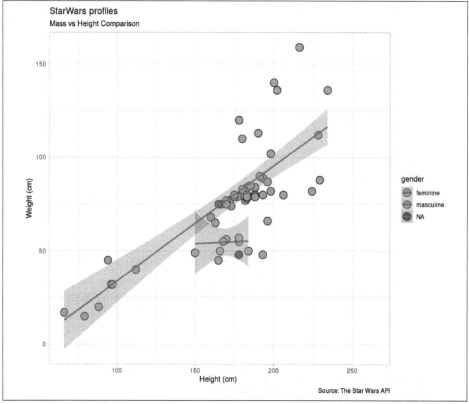

Figure 5-3. An advanced ggplot2 plot

Now that we covered static visualizations, let's see how to make them more interesting by adding interactivity!

Interactive Visualizations

Interactivity can be a great aid to exploratory plots. Two excellent R packages stand out: Leaflet (*https://rstudio.github.io/leaflet*) and plotly (*https://plotly.com*).

 Interactivity in Python and R is often based on an underlying Java-Script codebase. Packages like Leaflet and plotly make this work easier by providing a higher-level interface. Low-level packages for interactive graphics, like D3.js (*https://d3js.org*), can be overwhelming to learn for the novice. Thus, we'd encourage learning a high-level framework, such as dimple.js (*http://dimplejs.org*) instead.

Different datasets require different visualization methods. We covered the case of a standard tabular dataset (`starwars`), but how about something different? We'll have a go at visualizing data with a spatial dimension and use it to show R's excellent capabilities in producing interactive plots. For this, I selected the Shared Cars Locations dataset (*https://oreil.ly/FCKLG*). It provides the locations of car-sharing vehicles in Tel-Aviv, Israel. Can we show those on a map?

```
library(leaflet)
leaflet(data = shared_cars_data[1:20, ]) %>%
        addTiles() %>%
        addMarkers(lng = longitude, lat = latitude)
```

In this case, we subset the data using the first 20 rows only (to make the visualization less cluttered). The `addTiles` function provides the map background, with the street and city names.[6] The next step is to add the markers that specify the car locations by using `addMarkers`. The result of this relatively simple operation is shown in Figure 5-4.

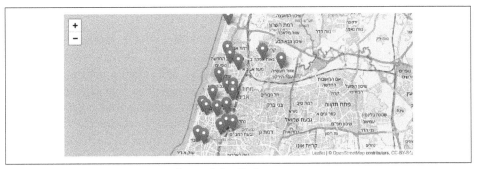

Figure 5-4. An interactive map plot with leaflet[7]

---

6 Explore the official documentation (*https://rstudio.github.io/leaflet*) for different map styles.

7 A color version is available for print readers online (*https://oreil.ly/prmd_5-4*).

As with the best data science tools, packages like Leaflet hide a lot of complexity under the hood. They do much of the heavy lifting necessary for advanced visualization and enable the data scientist to do what they do best—focus on the data. There are many more advanced features available in Leaflet, and we encourage the motivated user to explore them.

 As our book's subtitle suggests, we are always attempting to take the best of both worlds. So one easy way to do it is to use the ggplotly command from the plotly package and pass it a ggplot2 plot. This will make the plot interactive!

Hopefully, this section has made clear why the EDA workflow makes using R and tools such as ggplot2 and Leaflet the best options. We've just scratched the surface on what's possible, and if one decides to go deeper into the data visualization aspects, there are a ton of great resources available.

Machine Learning

Nowadays, data science is used almost synonymously with machine learning (ML). While there are many different workflows necessary for a data science project (Figure 5-1), ML often steals the focus of aspiring data scientists. This is partly due to an increasing growth surge in recent years due to the availability of large amounts of data, better computing resources (such as better CPUs and GPUs), and the need for predictions and automation in modern business. In the early days of the field, it was known under a different name—statistical learning. As previously mentioned, statistics has been historically the primary domain of R. Thus there were good tools available early on for doing ML in it. However, this has changed in recent years, and Python's tools have mostly overtaken its statistical competitor.

One can trace Python's ML ecosystem's success to one specific package: scikit-learn (*https://oreil.ly/ZCR55*). Since its early versions, the core development team has focused on designing an accessible and easy-to-use API. They supported this with some of the most complete and accessible documentation available in the open source world. It's not only a reference documentation but also contains excellent tutorials on various specific modern ML applications, such as working with text data (*https://oreil.ly/ZXsrG*). scikit-learn provides access to almost all common ML algorithms out of the box.[8]

Let's have a look at some proof of why scikit-learn is so great for ML. First, we can demonstrate the model imports:

---

8 An overview of those is available on scikit-learn (*https://oreil.ly/6g26U*).

```
from sklearn.ensemble import RandomForestClassifier
from sklearn.tree import DecisionTreeClassifier
from sklearn.linear_model import LinearRegression
```

Here we can already see how consistently those models are designed—similar to books in a well-organized library; everything is in the right place. ML algorithms in scikit-learn are grouped based on their similarities. In this example, tree-based methods such as `DecisionTreeClassifier` belong to the `tree` module. In contrast, linear algorithms can be found in the `linear_model` one (i.e., if you want to perform a least absolute shrinkage and selection operator [LASSO] model, you can, as you might predict, find it in `linear_model.Lasso`). Such hierarchical design makes it easier to focus on writing code and not to search for documentation since any good autocomplete engine will find the relevant model for you.

 We discussed modules in Chapter 3, but it's a concept that bears repeating since it might be confusing for some R users. Modules in Python are nothing more than collections of organized scripts (based on some similarities, such as "data_processing"), which allows them to be imported into your applications, improving readability and making the codebase more organized.

Next, we need to prepare the data for modeling. An essential element of any ML project is splitting the data into train and test sets. While newer R packages such as mlr improve on this as well, scikit-learn has better (in terms of both consistency and syntax) functions available:

```
from sklearn.model_selection import train_test_split
X_train, X_test, y_train, y_test = train_test_split(X, y,
                                            test_size=0.33,
                                            random_state=42)
```

Suppose we have been consistent in the steps before and have followed traditional ML convention. In that case, we have the X object to store our features and 'y'—the labels (in the case of a supervised learning problem).[9] In this case, the data will be randomly split. The official way to do this in R's mlr is:

```
train_set = sample(task$nrow, 0.8 * task$nrow)
test_set = setdiff(seq_len(task$nrow), train_set)
```

This can be harder to understand, and there's little documentation on how to perform a more advanced split, such as by stratification, and another package might be required, increasing the learning curve and cognitive load on the data scientist. scikit-

---

9 For those readers new to ML, supervised learning is concerned with prediction tasks where a target is available (label), as compared to unsupervised learning where it's missing, and the prediction task is on uncovering groups in the data.

learn, on the other hand, provides a handy function in `StratifiedShuffleSplit`. The capabilities only increase further when we start to perform the actual modeling:

```
model = RandomForestClassifier()
model.fit(X_train, y_train)
predictions = model.predict(X_test)
```

These three code lines are all we need to initialize the model with default parameters, fit (train) it on the training dataset, and predict on the test one. This pattern is consistent across projects (except for the model initialization, where one selects their algorithm of choice and its parameters—those do differ, of course). A visual comparison between several different packages (from other developers and purposes) is shown in Figure 5-6. Finally, let's compute some performance metrics; many of them are handily available:

```
from sklearn import metrics

acc = metrics.accuracy_score(predictions, y_test)
conf_matrix = metrics.confusion_matrix(predictions, y_test)
classif_report = metrics.classification_report(predictions, y_test)
```

The `metrics` module contains everything needed to check our model's performance with a simple and predictable API. The pattern of `fit` and `predict` we saw earlier has been so influential in the open source world that it has been widely adopted by other packages, such as Yellowbrick (a package for model performance visualization):

```
from yellowbrick.regressor import ResidualsPlot

visualizer = ResidualsPlot(regr)

visualizer.fit(X_train, y_train)
visualizer.score(X_test, y_test)
visualizer.show()
```

There are many visualizations available in Yellowbrick (Figure 5-5), all obtained with a similar procedure. The consistency and ease of use are among the significant reasons users want to use Python for ML. It enables the user to focus on the task at hand and not on writing code and sifting through tedious documentation pages. There were changes in R packages in recent years aiming at reducing those deficiencies. Such packages most notably include mlr and tidymodels. Still, they are not widely used, but perhaps this pattern can change in the future. There is an additional factor to consider here, which is similar to the ecosystem interoperability we saw in Chapter 4. scikit-learn works very well with other tools in Python that are necessary for the development and deployment of ML models. Such tools include database connections, high-performance computing packages, testing frameworks, and deployment frameworks. Writing the ML code in scikit-learn will enable the data scientist to be a more productive part of a data team (just imagine the expression on your data engineering colleagues' faces when you deliver an mlr model to them for deployment).

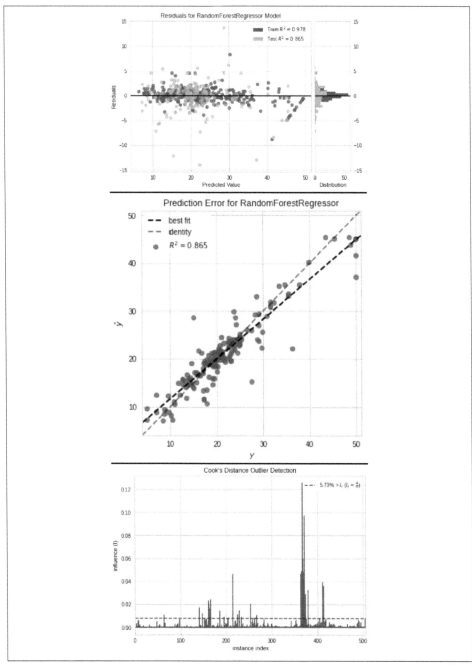

Figure 5-5. Different possible Yellowbrick regression plots

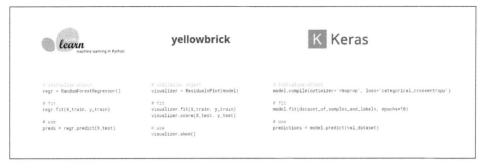

Figure 5-6. API consistency overview in the Python ML ecosystem

---

Deep Learning

We won't cover deep learning (DL) extensively here since most of the rationale from scikit-learn (and Python in general) applies to it as well. Still, due to its increasing importance in modern data science, it deserves a few additional comments.

The DL workflow has been mostly supported by two competing open source frameworks: TensorFlow (*https://www.tensorflow.org*) (from Google) and PyTorch (*https://pytorch.org*) (from Facebook). There is an additional framework, which was eventually included in TensorFlow, called Keras (*https://keras.io*). It provides a higher level of abstraction API to the TensorFlow functions, lowering the learning curve. There have been two notable developments in the R ecosystem regarding those DL frameworks. TensorFlow and Keras have been ported by using the reticulate package (*https://rstudio.github.io/reticulate*) (we'll cover it in Chapter 6), which calls Python under the hood. PyTorch, on the other hand, has been faithfully re-created on top of `libtorch`, the C++ backend of PyTorch in the torch package (*https://torch.mlverse.org*).

Due to those points, our recommendation is to use the Python tools for a DL workflow, based on Keras and TensorFlow, except using torch, in the case you have an existing R codebase.

To wrap up this section, we can summarize the main points about the ML workflow and why Python tools better support it:

1. Focus has moved to real-time predictions and automation.
2. The Python ML workflow provides a more consistent and easy-to-use API.
3. Python is more of a glue language, ideal for combining different software components (i.e., frontend/backend and databases).[10]

---

10 For a visual of ML architectures' complexity, have a look at Google's MLOps document (*https://oreil.ly/SIGu4*).

In the next section, we'll go deeper into the third part of this list and demonstrate the recommended data engineering workflow.

Data Engineering

Despite the ML tools' advancements in recent years, the completion rate of such projects in companies remains low. One reason that is often credited for this is the lack of data engineering (DE) support. To apply ML and advanced analytics, companies need the infrastructural foundation provided by data engineers, including databases, data processing pipelines, testing, and deployment tools. Of course, this forms a separate job title—data engineer. Still, data scientists need to interface with (and sometimes implement themselves) those technologies to ensure data science projects are completed successfully.

While DE is a massive field, we'll focus on a subset for this section. We selected model deployment for this since it's the most common DE workflow that a data scientist might need to participate in. So what is ML deployment? Most of the time, this means creating an application programming interface (API) and making it available to other applications, either internally or externally (to customers, this is called "exposing" an API, to be "consumed"). Commonly ML models are deployed via a representational state transfer (REST) interface.[11]

ML model deployment, compared to the other topics in this chapter, requires interfacing with many different technologies not directly related to data science. These include web frameworks, CSS, HTML, JavaScript, cloud servers, load balancers, and others. Thus it's not surprising that Python tools dominate here—as we covered before, it's a fantastic glue language.[12]

 The model deployment workflow requires code to be executed on other machines rather than the local one where the data scientist performs their daily work. This hits the "it works on my machine" problem right on the head. There are different ways to deal with managing different environments consistently, ranging from simple to complex. A simple way to do this is to use a *requirements.txt* file, where all dependencies are specified. A more complex option, which is often used in large-scale, critical deployments, uses container solutions such as Docker (*https://www.docker.com*). This dependency management is much easier to achieve in Python than in R.

---

11 To learn more about what is REST, have a look at Wikipedia's page (*https://oreil.ly/OPaFE*).

12 The R alternative to Flask is `plumber`. The RStudio IDE provides a friendly interface to use this tool, but still, it is lagging in options and adoption in the ML community.

One of the most popular tools to create an API is Python's Flask (*https://oreil.ly/6S7ag*)—a micro-framework (*https://oreil.ly/SSs59*). It provides a minimalist interface that is easy to extend with other tools, such as ones providing user authentication or better design. To get started, we'll go through a small example. We would need a typical Python installation with some other additional configurations such as a virtual environment and a GUI to query the API.[13] Let's get started!

> Recently competitors to Flask have sprung up. They serve the same purpose but with an increased focus on ML. Two popular examples include BentoML (*https://www.bentoml.ai*) and FastAPI (*https://FastAPI.tiangolo.com*). Those frameworks provide you with some additional options that make ML deployment easier. Remember that Flask was initially built for web development APIs, and the needs of an ML project can be different.

We'll be building an API that predicts housing prices.[14] It's always prudent to start with the end goal in mind and how we'd like such a predictive model to be used by an external application or an end user. In this case, we can imagine our API to be integrated into an online house rental portal.

For brevity, we'll omit the model training part. Imagine that you have followed a traditional scikit-learn model development. The results of the predictive model are stored in a .pkl (Pickle object, the standard Python way to store objects on disk). This process is called *serialization*, and we need to do it to use the model in the API later:

```
import pickle

# model preparation and training part
# ...

# model serialization
outfile = open("models/regr.pkl", "wb")
pickle.dump(regr, outfile)
outfile.close()

print("Model trained & stored!")
```

---

13 For brevity, we will not go deeper into setting up virtual environments here. We urge the dedicated reader to read up upon the virtualenv (*https://oreil.ly/oqn0Y*) and renv (*https://oreil.ly/7szIG*) tools, covered in Chapter 3.

14 The dataset is "Boston Housing," available on scikit-learn (*https://oreil.ly/xpTI6*).

We can save this code in a script called *train_model.py*. By running it: `python train_model.py`, the pickled model will be produced and saved. Figure 5-7 provides an overview of how the different components fit.

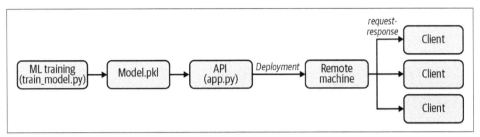

Figure 5-7. Example architecture for an ML API

 In our example, the API provides just one functionality—the ability to predict a housing price on a dataset. Often in the real world, the same application might need to do different things. This is organized by creating different endpoints. For example, there might be an endpoint for triggering a data preparation script and a separate inference one.

Let's use Flask next:

```python
import pickle
import numpy as np
from ast import literal_eval ❶
from flask import Flask, request, jsonify

app = Flask(__name__) ❷

infile = open("models/regr.pkl", "rb") ❸
regr = pickle.load(infile)
infile.close()

@app.route('/') ❹
def predict(methods=["GET"]):
    payload = request.json["data"]
    input_data = np.array(literal_eval(payload)).reshape(1, -1)
    prediction = regr.predict(input_data) ❺

    return jsonify({
        "prediction": round(float(prediction), 3) ❻
    })


if __name__ == '__main__':
    app.run(debug=True)
```

❶ We use this function to specify that the payload string object is actually a dictionary.

❷ Here we create an object that holds the app.

❸ In these several lines we load the serialized model.

❹ This Python decorator creates an endpoint.

❺ At this step, the serialized model is used for inference.

❻ The inference results are returned in a JSON format.

This code is added to a file *app.py*. Once you execute this script, the command line will output a local URL. We can then use a tool such as Postman to query it.[15] Have a look at Figure 5-8 to see how such a query works. Voilà—we built an ML API!

Figure 5-8. Querying the ML API with Postman

---

15 If you are more of a command-line person, have a look at `curl`.

> ## Cloud Deployment
>
> After you're done with writing and testing the ML API code, the next phase would be to deploy it. Of course, you could use your computer as a server and expose it to the internet, but you can imagine that doesn't scale very well (you have to keep your machine running, and it might run out of resources). One of the significant changes seen in recent years in terms of DE tools is the advent of cloud computing. Cloud platforms such as AWS or Google Cloud Provider (GCP) provide you with excellent opportunities and deploy your apps. Your Flask API can be deployed via a cloud service such as AWS Elastic Beanstalk (*https://oreil.ly/emrdf*) or Google App Engine (*https://oreil.ly/bgPHa*).

Due to the "glue-like" nature of Python packages, they dominate the DE workflow. If a data scientist can write such applications on their own in Python, the success of the complete data project is ensured.

Reporting

Every data scientist is aware (perhaps painfully so) of how vital communication is for their daily work. It's also an often underrated skill, so this mantra bears repeating. So, what is more important than one of the essential deliverables of a data science project —reporting your results?

There are different reporting methods available. The most typical use case for a data scientist is to create a document, or a slide deck, containing the results of the analysis they have performed on a dataset. This is usually a collection of visualizations with an associated text and a consistent storyline (i.e., going through the different stages of a project life cycle—data importing, cleaning, and visualization). There are other situations where the report has to be referred to often and updated in real time, called *dashboards*. And finally, some reports allow the end user to explore them more interactively. We'll go through those three report types in the following subsections.

Static Reporting

The popularization of the markdown language helps data scientists focus on writing code and associated thoughts instead of the tool itself. A flavor of this language—R Markdown (RMD) is widely used in the R community. This allows for the concept of *literate programming*, where the code is mixed with the analysis. The RStudio IDE provides even further functionality with tools such as R Notebooks (*https://oreil.ly/3STa2*). This is how easy writing an RMD report is:

```
# Analyzing Star Wars

First we start by importing the data.
```

```{r}
library(dplyr)

data(starwars)
```

Then we can have a look at the result.

This *.rmd* file can then be *knit* (compiled) into a *.pdf* or an *.html* (best for interactive plots), creating a beautiful report. There are additional templates to create even slides, dashboards, and websites from RMD files. Have a look at Figure 5-9 to check it out in action.

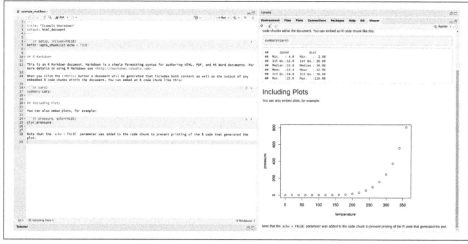

Figure 5-9. R Markdown editing within RStudio

As with everything in the open source world, data scientists worldwide have contributed to the further development of RMD. There are many templates available for RMD, enabling users to create everything from a custom-styled report to a dynamically generated blogging website.

 The widely adopted alternative to RMD in the Python world is the Jupyter Notebook (*https://jupyter.org*) (along with its newer version: Jupyter Lab (*https://oreil.ly/VgfmD*)). The "r" in Jupyter stands for R, and it is certainly possible to use that, but we argue that the RMD notebooks in RStudio provide a better interface, at least for R work.

Interactive Reporting

What if we want to be able to let the recipients of our report do some work as well? If we allow for some interactivity, our end users would answer questions for themselves without relying on us to go back, change the code, and regenerate the graphs. There are several tools available,[16] but most of them pale in comparison to the ease of use and capabilities of R's shiny package.[17]

Using this package requires a bit of a different way of writing R code, but you will create fantastic applications once you get used to it. Let's go through a basic yet practical example. shiny apps consist of two fundamental elements: the user interface (UI) and the server logic. Those are often even separated into two files. For simplicity we'll use the single file layout and use two functions for the app:

```
library(shiny)

ui <- fluidPage( ❶

    titlePanel("StarWars Characters"),

    sidebarLayout(
        sidebarPanel(
            numericInput("height", "Minimum Height:", 0, min = 1, max = 1000), ❷
            numericInput("weight", "Minimum Weight:", 0, min = 1, max = 1000),
            hr(),
            helpText("Data from dplyr package.")
        ),

        mainPanel(
            plotOutput("distPlot") ❸
        )
    )
)
```

❶ This function specifies that we want to have a "fluid" layout that makes the app *responsive*—easy to read on a variety of devices, such as smartphones.

❷ Add the dynamic input for the user.

❸ Add a dedicated area for the output.

---

16 There's an advanced new tool in Python, called Streamlit (*https://www.streamlit.io*), but it is yet to gain in popularity and adoption.

17 To get inspired with what's possible in shiny, look at the gallery of use cases at the RStudio website (*https://oreil.ly/DVbDd*).

The ui object contains all the frontend parts of the application. The actual computation happens in the following function; we'll be adding the ggplot from the DV section:

```r
server <- function(input, output) { ❶

    output$distPlot <- renderPlot({ ❷
        starwars_filtered <- starwars %>%
            filter(height > input$height & mass > input$weight) ❸
        ggplot(starwars_filtered, aes(x = height, y = mass, fill = gender)) +
            geom_point(pch = 21, size = 5) +
            theme_light() +
            geom_smooth(method = "lm") +
            labs(x = "Height", y = "Mass",
            title = "StarWars Characters Mass vs Height Comparison",
            subtitle = "Each dot represents a separate character",
            caption = "Data Source: starwars (dplyr)") ❹
    })
}
```

❶ The server needs two things: input and output.

❷ There is just one output in our case.

❸ We can add all kinds of R computations here, as in a regular R script.

❹ The most recent item (in this case, a plot) is returned for display in the frontend.

The computation happens in this function. In the end, we need to pass those two functions to shinyApp to see it in action. This step initiates the shiny backend, which supports the inputs on from the ui function with computations in the server one. The results of this are shown in Figure 5-10.

```r
shinyApp(ui = ui, server = server)
```

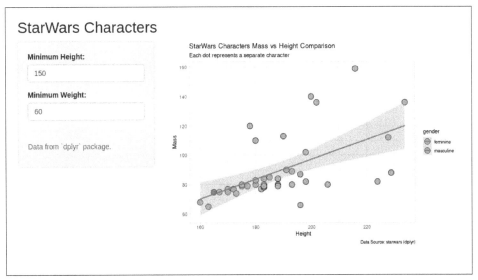

Figure 5-10. An interactive report with shiny[18]

One difference for our shiny app that might make it trickier to use than our markdown files is that you would need to host the application on a remote machine. For a normal *.rmd* on the files, you need to knit the file into a PDF and then share it. How such applications are deployed is beyond this book's scope.

Creating reports is a small but vital component of data science work. This is how your work is shown to the outside world, be it your manager or another department. Even if you have done a great job in your analysis, it will often be judged by how well you communicate the process and results. Tools of literate programming such as RMD and more advanced interactive reports in shiny can go a long way to creating state-of-the-art reports. In the final chapter of this book, Chapter 7, we'll provide a great example of this in action.

Final Thoughts

In this chapter, we went through the most essential workflows in a data science project and discovered the best tools in R and Python. In terms of EDA and reporting, R can be crowned the king. Packages such as ggplot2 are peerless in the data science community, and shiny can allow for fascinating new ways to present data science results to stakeholders and colleagues. In the ML and DE worlds, Python's glue-like nature provides fantastic options, enabling modern data scientists to focus on the work rather than the tools.

---

18 A color version is available for print readers online (*https://oreil.ly/prmd_5-10*).

## PART IV

Bilingualism III: Becoming Synergistic

So far in this book we have been exploring the two languages in quite isolated cases. We learned how to learn one from the other, and in which data formats and work-flows they excel. In all of those cases the separation has been quite distinct—for some tasks R excels, and in others its general-purpose counterpart Python is a better choice.

Chapter 6

In this chapter we'll take a different perspective—one that can be heralding a new way to work with programming languages in the future.

Chapter 7

As the final chapter of the book, it is fitting to apply all the concepts we have learned so far. We'll do this by going through a real-world case study of bilingual data science.

Using the Two Languages Synergistically

Rick J. Scavetta

Interoperability, the ability for different programming languages to work together, is a cornerstone of computing. Ideally objects can be shared directly between the two languages. As you can imagine, this can be problematic for a variety of reasons, like memory usage and incompatible data storage structures to name just two. Although there have been several attempts to implement this smoothly between Python and R, it's only been in the past couple of years that a reasonably functional kit had come to fruition. I'll discuss this in "Interoperability" on page 120. But it's useful to first return to the basics. This will not only give context to appreciate smooth interoperability later on but also give you a basic solution that may already meet your needs. Nonetheless, if you want to get started with interoperability, you can skip the next section.

Faux Operability

The most basic type of interoperability, sometimes called cross-talk, is more of a faux operability. Here, we execute predefined scripts across languages, passing information between them using files as intermediaries. Imagine the following situation, which I've diagrammed in Figure 6-1.

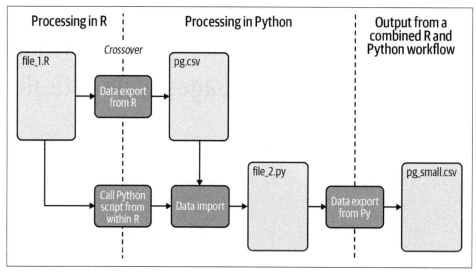

Figure 6-1. An example of cross-talk to facilitate interoperability

In R, after performing some necessary work on an object, e.g., `PlantGrowth`, we execute:

```
# (Previous interesting and complicated steps omitted)

# Write a data.frame of interest to a file ...
rio::export(PlantGrowth, "pg.csv")

# ... which is then processed by a Python script
system('~/.venv/bin/python3 myScript_2.py < "pg.csv"')
```

The `system()` function executes a system command, provided as a character argument. The command is made up of four parts.

First, *~/.venv/bin/python3* is the location of the Python executable within our virtual environment, assuming that you've created one. We could have also included this in the script's *shebang* first line as `#!/.venv/bin/env python3`. This ensures that the script is executed in the environment in which it was created. See "Virtual Environments" on page 50 if this sounds strange to you.

Second, *myScript_2.py* is the name of the Python file that contains the commands we want to execute.

Third, `<` allows us to redirect `stdin` from the rhs to the file on the lhs.[1]

---

1 Recall that rhs is the *right-hand side* and lhs is the *left-hand side* when calling operators, in this case `<`.

Fourth, "*pg.csv*" is the `stdin`. You may recall that there are three standard *channels*, or *streams*, for command-line functions. `stdin` for the *standard input*, `stdout` for the *standard output*, and `stderr` for the standard error. Here, `stdin` is hardcoded. It's a character string that corresponds to a file: "*pg.csv*," which was exported in the previous command. Hardcoding should be avoided for the most part, and we're sure you can imagine many ways to make this dynamic. That's not really our focus here; the point is to feed some input into a Python script.

Thus, we're executing a Python script that takes `stdin` from within an R script, and that `stdin` in itself is a product of the R script. Let's take a look at the minimal components of this Python script:

```python
import sys
import pandas as pd

# import the file specified by the standard input
myFile = pd.read_csv(sys.stdin)

# (Fantastically complex and very Pythonic code omitted)

# Write the first four lines to a file
myFile.head(4).to_csv("pg_small.csv")
```

First we need the `sys` module to handle `stdin` (`sys.stdin`). We import the file, represented by `sys.stdin` using pandas, and after our Python script works its magic we export some other output using the `to_csv()` method.

There are a lot of things wrong with this method, and we'll get to them soon. But the point is that it works, and sometimes, it's exactly what you need. Working in a research laboratory, I often had to provide results to colleagues quickly. I mean this literally, since very expensive cell cultures would die and a week's worth of work would be wasted if the results were not ready. Preprocessing of proprietary raw data and access to a secure server prohibited my colleagues from executing automated R scripts. My solution was to first process the machine-generated proprietary data with software specialized for the task. Then I was able to use a macOS Automator service to execute a Perl script on that output, which was now my `stdin`. This Perl script then called an R script that produced a file of a plot with all the relevant information clearly displayed in the title. It wasn't the most open or elegant solution, but it worked and I got my plots with one mouse click in about half a second without any extra websites or logins. Life was good, so what's the problem?

Well, there are several problems. Let's consider three.

First, in retrospect, I could have probably executed the entire workflow in R (excluding the proprietary preprocessing). It's necessary to consider simplifying a workflow and having a good reason to use multiple languages. Deciding when and why to combine Python and R has come up throughout this book.

Second, there are a lot of moving parts. We have several files and we're even producing additional intermediate ones. This increases the chance for error and confusion. That's not terrible, but we better take care to keep things organized.

Third, in many cases, this workflow works well when we can export an R *data.frame* as a CSV file, which pandas can easily import. For more complex data structures, you can export one or more R objects as an RDATA or RDS format file. The python pyreadr package provides functions to import these files and provide access to each object stored in a `dict`.

Cross-talk is great, but true interoperability smooths out the wrinkles in this process quite nicely. There are two widely used frameworks; the choice of which to use will depend on which language is your starting point.

Interoperability

If you're primarily using R and want access to Python, then the R package reticulate is the way to go. Conversely, if you're primarily using Python and want access to R, then the Python module rpy2 is the tool for you. We can summarize this in Table 6-1 and Table 6-2.[2] In each table, read each line as a sentence beginning with the column headers.

Table 6-1. Interoperability granted by reticulate

Access	Using command
Python Functions	in R, `pd <- library(pandas); pd$read_csv()`
Python Objects	in R, `py$objName`
R Objects	in Python, `r.objName`

Table 6-2. Interoperability granted by rpy2 when writing in Python

Access	Using command
R Functions	in Python, `import rpy2.robjects.lib.ggplot2 as ggplot2`
R Packages	in Python, `r_cluster = importr('cluster')`
R Objects	in Python, `foo_py = robjects.r['foo_r']`

The commands in Table 6-1 and Table 6-2 reveal how to access all varieties of objects from one language directly from the other. In addition, we can also directly call functions. This is a real milestone since it relieves us of having to force one language to do tasks that it doesn't excel at and means that we don't need to reinvent the wheel,

---

2 In these tables we make a distinction between functions and objects. Recall that functions are themselves just objects, but we don't need to worry about these details at the moment.

---

introducing redundancy between the languages. At the time of writing, it was not possible to access R functions from within Python in reticulate. You may attempt to use reticulate for this task, but it would be easier to pass an object back to R and execute R commands natively.

reticulate first appeared on CRAN in 2017, and has recently gained in popularity as it matured. This package is developed by RStudio and is well integrated into the RStudio IDE itself, which is pretty convenient. However, at the time of writing, there are some troublesome features (bugs?) that require some finesse (see the warning box that follows). A good first step is to ensure you are using the latest public release of RStudio and the latest version of the reticulate package and any associated packages, such as knitr.

reticulate is well supported and stable enough to be used in production. Nonetheless, you may encounter issues depending on your system and software versions. Since this tool combines technologies, it can also be difficult to debug, and documentation is still somewhat scarce. Stay up-to-date with new versions as they are released. If you encounter issues on your local machine, call up our RStudio Cloud (*https://rstudio.cloud/project/2534578*) project.

In this section we'll begin with two scripts, listed in Table 6-3. You'll find these in this chapter's folder in the book's repository (*https://github.com/moderndatadesign/ PyR4MDS*).

Table 6-3. Up and running with reticulate

File	Description
0 - Setup.Rmd	Setting up reticulate and virtual environments
1 - Activate.R	Activating a Python virtual environment

Let's begin with the R script, 0 - Setup.R. Make sure you've installed reticulate and have initialized it in your environment:

```
library(reticulate)
```

First, we need to specify which build of Python we'll use. You can let R use your system default or set the specific build of Python you want to use by going to Tools > "Project options" and selecting the Python icon (see Figure 6-2).

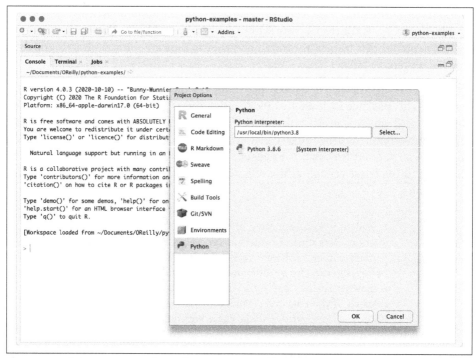

Figure 6-2. Selecting a specific Python version and build to use

Let's check to see the version we're using:

```
reticulate::py_config()
```

```
python:        /usr/local/bin/python3.8
libpython:     /Library/Frameworks/Python.framework/Versions/3.8...
pythonhome:    /Library/Frameworks/Python.framework/Versions/3.8...
version:       3.8.6 (v3.8.6:db455296be, Sep 23 2020, 13:31:39) ...
numpy:         [NOT FOUND]
sys:           [builtin module]
```

To be clear, we don't need to use RStudio to set the Python version. This is just a convenience feature. We could have executed:

```
use_python("/usr/local/bin/python3.8", required = TRUE)
```

Note that this function just makes a *suggestion* and doesn't result in an error if the desired build is not found unless the `required` argument is set to `TRUE`.

Before we proceed, we'll want to establish a virtual environment. If you're on Windows, you'll have to use a `conda` environment, which we'll get to in a minute. For everyone else, use the following command to create a virtual environment called `modern_data`:

```
virtualenv_create("modern_data")
```

Previously, when we used the venv package in Python, the virtual environment was stored as a hidden directory (typically called *.venv* in the project directory). So where are the Python virtual environments now? We can have a look with the following command:

```
virtualenv_root()

[1] "~/.virtualenvs"
```

They are all stored in a hidden folder in the *root* directory. We can see all our virtual environment using the following command:

```
virtualenv_list()
```

 As is the case for most popular data science packages, there is a cheat sheet available for reticulate. You can download it directly from RStudio (*https://raw.githubusercontent.com/rstudio/cheat sheets/master/reticulate.pdf*).

This is a departure from what we saw with virtual environments in Python, where they were stored within the project directory. Nonetheless, it's convenient since we can easily reuse a good environment for many projects.

Note that to remove a virtual environment, we need to pass the path as such:

```
virtualenv_remove("~/modern_data")
```

The next step is to install the appropriate packages:

```
virtualenv_install("modern_data", "pandas")
```

Alternatively, you can use the Tidyverse purrr::map() function to install many packages:

```
library(tidyverse)
c("scikit-learn", "pandas", "seaborn") %>%
  map(~ virtualenv_install("modern_data", .))
```

If you're on Windows, use the following commands:

```
# For windows users:
# Install a minimal version of Conda
install_miniconda()

# List your Conda virtual environments
conda_list()

# Create a new virtual environment
conda_create("modern_data")

# Install a single...
conda_install("modern_data", "scikit-learn")
```

```
#...or multiple packages:
library(tidyverse)
c("scikit-learn", "pandas", "seaborn") %>%
  map(~ conda_install("modern_data", .))
```

The final step is to activate our virtual environment. This seems to be an area under rapid development. Different error messages, or none at all, are produced depending on your versions of reticulate and RStudio, making them harder to debug. From my experience, your safest bet is to (i) make sure all your R packages, as well as RStudio, are up-to-date and (ii) restart R before activating your virtual environment. You can do this in the RStudio menu Session > Restart R, the keyboard shortcut Shift-Command/Ctrl + F10, or executing the command **.rs.restartR()**. You can also literally close and restart RStudio. This ensures that there is no Python build in active use and we can establish one from scratch. Thus, we have one R script for set up, where we create a virtual environment and install packages, and another with our actual analysis, where we load reticulate and activate our virtual environment.

```
library(reticulate)
use_virtualenv("modern_data", required = TRUE)

# Alternatively, for miniconda:
# use_miniconda("modern_data")
```

And finally, we can confirm which build we have using:

```
py_config()
```

You should see the following output. Importantly, make sure that the path to your virtual environment is stated in the first line: /.virtualenvs/modern_data/bin/python.

```
python:          /Users/user_name/.virtualenvs/modern_data/bin/python
libpython:       /Library/Frameworks/Python.framework/Versions/3.8...
pythonhome:      /Users/user_name/.virtualenvs/modern_data...
version:         3.8.6 (v3.8.6:db455296be, Sep 23 2020, 13:31:39)
numpy:           /Users/user_name/.virtualenvs/modern_data/lib/python3.8/...
numpy_version:   1.20.1
```

If you see something like /usr/local/bin/python3.8, then RStudio is still directed to use the Python version you defined at the beginning of the chapter and not a virtual environment. This may serve you well, but it is preferable to use a virtual environment.

Going Deeper

At this point, we've created a virtual environment, installed some packages in it, restarted R, and activated the virtual environment. These steps are covered in the the scripts 0 - Setup.R and 1 - Activate.R. For the rest of this section I'll cover ways to pass information between R and Python, which I've summarized in Table 6-4.

Table 6-4. Interoperability granted by reticulate

File	Description
2 - Passing objects.Rmd	Pass objects between R and Python in an R Markdown document
3 - Using functions.Rmd	Call Python in an R Markdown document
4 - Calling scripts.Rmd	Call Python by sourcing a Python script
5 - Interactive mode.R	Call Python using a Python REPL console
6 - Interactive document.Rmd	Call Python with dynamic input in an interactive document

> Why "reticulate"? The reticulated python is a species of python found in Southeast Asia. They are the world's longest snakes and longest reptiles. The species name, *Malayopython reticulatus*, is Latin meaning "net-like," or reticulated, and is a reference to the complex color pattern.

I'll consider the scenarios in Table 6-1 in the following subsections. To follow along with these examples, please ensure that you have followed the set up and activation instructions found in 0 - Setup.R and 1 - Activate.R—both in the book's code repository (*https://github.com/moderndatadesign/PyR4MDS*). You'll need to have the modern_data virtual environment and the preceding list of packages installed. If you're using Miniconda, be sure to use the correct command given in each file to activate your virtual environment.

Pass Objects Between R and Python in an R Markdown Document

The following commands can be found in the file *2 - Passing objects.Rmd*. To access an R object in Python, use the r object, and to access a Python object in R, use the py object. Consider the following chunks found in an R Markdown document:

```
```{python}
a = 3.14
a
```

```{r}
py$a
```
```

The python object `a` is accessed in the R object `py` using the $ notation. In the opposite direction:

```{r}
b <- 42
b
```

```{python}
r.b
```

In Python, call the `r` object and use `.` notation to access R objects by name. These are scalars, or simple vectors, but of course we can pass more complex items directly between the two languages. reticulate will take care of object conversion for us. Consider the following case:

```{r}
# A built-in data frame
head(PlantGrowth)
```

```{python}
r.PlantGrowth.head()
```

An R `data.frame` is accessed as a Python `pd.DataFrame`. However, if you don't have pandas installed you'll see a `dict` object, a Python dictionary.

A Python NumPy `ndarray` will be converted to an R `matrix`:[3]

```{python eval = TRUE}
from sklearn.datasets import load_iris

iris = load_iris()
iris.data[:6]
```

A Python NumPy `ndarray` as an R `matrix`:

```{r eval = TRUE}
head(py$iris$data)
```

Notice how the `.` notation in Python, `iris.data`, is automatically accessible using the $ notation in R: `py$iris$data`. This holds true for nested objects, methods, and attributes, just as they would in Python.

---

3 Refer to the Appendix for a summary of data structures.

Call Python in an R Markdown Document

The following commands can be found in the file *3 - Using functions.Rmd*. We'll continue to use the classic iris dataset that we accessed in Python in the previous section. Inside an R Markdown document, we'll access a Python function, which allows us to access the trained support vector machine classifier to predict classification on new values. This is a very naïve machine learning workflow and is not intended to produce a valuable model. The point is to demonstrate how to access a model from Python in R.

The entire model configuration is defined here:

```{python}
# import modules
from sklearn import datasets
from sklearn.svm import SVC

# load the data:
iris = datasets.load_iris()

# Create an instance of the SVC, _Support Vector Classification_, class.
clf = SVC()

# Train the model by calling the fit method on the target data, using target names
clf.fit(iris.data, iris.target_names[iris.target])

# Predict the class of new values, here the first three
clf.predict(iris.data[:3])
```

The method `clf.predict()` takes an `ndarray` as input and returns the named classification. To access this function in R, we can once again use the py object, as in py `$clf$predict()`. The `iris` dataset in R is a `data.frame`, where the fifth column is the classification. We must convert this to a Python object using `r_to_py()`, in this case excluding the fifth column.

```{r}
py$clf$predict(r_to_py(iris[-5]))
```

Call Python by Sourcing a Python Script

The following commands can be found in the file *4 - Calling scripts.Rmd* and *4b - Calling scripts.R*. In this scenario we'll execute an entire Python script and access all objects and functions available therein. To do this we can call:

```
source_python("SVC_iris.py")
```

This works just as well in an R Markdown document as in a script.

Although this appears very similar to the previous section, there is a very important distinction. Python environments activated in this manner provide functions and objects directly. Thus we can call:

```
clf$predict(r_to_py(iris[-5]))
```

This is convenient, but also disconcerting. Not only has the syntax changed, i.e., no need for py$, but objects loaded in the R environment may conflict. Python objects will mask R objects, so be very careful about naming conflicts! You'll notice that in SVC_iris.py we've renamed the Python iris dataset to iris_py to avoid problems when calling iris in R.

Call Python Using the REPL

The following commands can be found in the file *5 - Interactive mode.R*. In this scenario we'll start up a Python REPL console, using the following command:

```
repl_python()
```

 REPL stands for read-eval-print loop. It is a common feature in many languages where the user can experiment in an interactive way, as opposed to writing a script that needs to be run.

This will allow you to directly execute Python commands in an interpreter. For example, try executing the commands we saw in the last example:

```
from sklearn import datasets
from sklearn.svm import SVC
iris = datasets.load_iris()
clf = SVC()
clf.fit(iris.data, iris.target_names[iris.target])
clf.predict(iris.data[:3])
```

We can exit the interpreter by executing the Python exit command:

```
exit
```

Just like we've seen before, the functions and objects in this Python environment can be accessed in R. This is truly interactive programming because we're executing commands directly in the console. Although we present this scenario for the sake of completeness, repl_python() is not really meant to be used in everyday practice. Actually, it's what is called when an R Markdown chunk uses a Python kernel. So although you can do this, be cautious! This presents a considerable problem in reproducibility and automation, but you may find it useful for quickly checking some commands.

Call Python with Dynamic Input in an Interactive Document

The following commands can be found in the file *6 - Interactive document.Rmd*.

By now we've seen all the core functionality of reticulate. Here we'll go beyond that and show a very simple way to introduce interactivity using a shiny runtime in an R Markdown document. To see the interactivity, make sure you have the shiny package installed and that you render the document to HTML. In RStudio, you can do this by clicking the Run Document button when the file is open.

First, in the header of our document we need to specify this new runtime environment:

```
---
title: "Python & R for the Modern Data Scientist"
subtitle: "A bilingual case study"
runtime: shiny
---
```

The following Python code, which we've seen, is executed in a Python chunk:

```
```{python}
from sklearn import datasets
from sklearn.svm import SVC
iris = datasets.load_iris()

clf = SVC()
clf.fit(iris.data, iris.target_names[iris.target])
```
```

In the final two chunks, we can use functions from the shiny package to produce a user interface. This consists of both input and output.

First, the input. We produce a slider for each of the four features using the `sliderInput()` function as follows to `sl`. The sliders for `sw`, `pl`, and `pw` are similar and can be found in the case study script.

```
sliderInput("sl", label = "Sepal length:",
            min = 4.3, max = 7.9, value = 4.5, step = 0.1)
```

Second, the output. We call the values from the sliders as named elements (`sl`, `sw`, `pl`, and `pw`) in an R list object called `input` (`input$sl`). These values are used as input to the Python `predict()` function. The Python output is assigned to the R object prediction:

```
prediction <- renderText({
  py$clf$predict(
    r_to_py(
      data.frame(
        sl = input$sl,
        sw = input$sw,
        pl = input$pl,
        pw = input$pw)
    )
  )
})
```

Finally, we call the R object `prediction` as an inline command, `` `r prediction` ``, to print the result to the screen as a sentence.

Final Thoughts

In this chapter we've covered the core components of the reticulate package, progressing from the essential set up to the basics and finally a simple yet powerful implementation that showcases the strengths of R, Python, and reticulate. Using this knowledge, we'll continue to a larger case study in the last chapter.

## CHAPTER 7

A Case Study in Bilingual Data Science

Rick J. Scavetta
Boyan Angelov

In this final chapter, our goal is to present a case study that demonstrates a sample of all the concepts and tools we've shown throughout this book. Although data science provides a practically overwhelming diversity of methods and applications, we typically rely on a core tool kit in our daily work. Thus, it's unlikely that you'll make use of *all* the tools presented in this book (or this case study, for that matter). But that's alright! We hope that you'll focus on those parts of the case study that are most relevant to your work and that you'll be inspired to be a modern, bilingual data scientist.

24 Years and 1.88 Million Wildfires

Our case study will focus on the US Wildfires dataset.[1] This dataset, released by the US Department of Agriculture (USDA), contains 1.88 million geo-referenced wildfire records. Collectively, these fires have resulted in the loss of 140 million acres of forest over 24 years. If you want to execute the code in this chapter, download the SQLite dataset from the USDA website (*https://doi.org/10.2737/RDS-2013-0009.4*) directly or from Kaggle (*https://oreil.ly/jyCsp*), and place it inside the *ch07/data* directory

There are 39 features, plus another shape variable in raw format. Many of these are unique identifiers or redundant categorical and continuous representations. Thus, to simplify our case study, we'll focus on a few features listed in Table 7-1.

---

1 Short, Karen C. 2017. Spatial wildfire occurrence data for the United States, 1992-2015, FPA_FOD_20170508. 4th Edition. Fort Collins, CO: Forest Service Research Data Archive. *https://doi.org/10.2737/RDS-2013-0009.4*.

Table 7-1. The Fires table contains 39 features describing over 1.88 million wildfires in the US from 1992 to 2015

Variable	Description
STAT_CAUSE_DESCR	Cause of the fire (the target variable)
OWNER_CODE	Code for primary owner of the land
DISCOVERY_DOY	Day of year of fire discovery or confirmation
FIRE_SIZE	Estimate of the final fire size (acres)
LATITUDE	Latitude (NAD83) of the fire
LONGITUDE	Longitude (NAD83) of the fire

We'll develop a classification model to predict the cause of a fire (STAT_CAUSE_CODE) using the five other features as features. The target and the model are secondary; this is not an ML case study. Thus, we're not going to focus on details such as cross-validation or hyperparameter tuning.[2] We'll also limit ourselves to observations from 2015 and exclude Hawaii and Alaska to reduce the dataset to a more manageable size. The end product of our case study will be to produce an interactive document that will allow us to input new predictor values, as depicted in Figure 7-1.[3]

Before we dig in, it's worth taking a moment to consider data lineage—from raw to product. Answering the following questions will help orientate us.

1. What is the end product?

2. How will it be used, and by whom?

3. Can we break down the project into component pieces?

4. How will each component be built? That is, Python or R? Which additional packages may be necessary?

5. How will these component pieces work together?

Answering these questions allows us to draw a path from the raw data to the end product, hopefully avoiding bottlenecks along the way. For Question 1, we've already stated that we want to build an interactive document. For the second question, to keep things simple, let's assume it's for us to easily input new feature values and see the model's prediction.

---

2 We'll leave a thorough development of a robust classification model to our motivated readers. Indeed, you may also be interested in a regression that predicts the final fire size in acres. Curious readers will note that a few interesting notebooks are available on Kaggle to get you started.

3 This is a far cry from developing, hosting, and deploying robust ML models, which, in any case, is not the focus of this book.

Questions 3–5 are what we've considered in this book. In Question 3, we imagine the parts as a series of steps for our overall workflow. Question 4 was addressed in Chapter 4 and Chapter 5. We summarize those steps in Table 7-2.

Table 7-2. The steps in our case study and their respective languages

Component/Step	Language	Additional packages?
1. Data importing	R	RSQLite, DBI
2. EDA and data visualization	R	ggplot2, GGally, visdat, naniar
3. Feature engineering	Python	scikit-learn
4. Machine learning	Python	scikit-learn
5. Mapping	R	Leaflet
6. Interactive web interface	R	shiny runtime in an R Markdown

Finally, Question 5 asks us to consider the project architecture. The diagram presented in Figure 7-1 shows how the steps in Table 7-2 will be linked together.

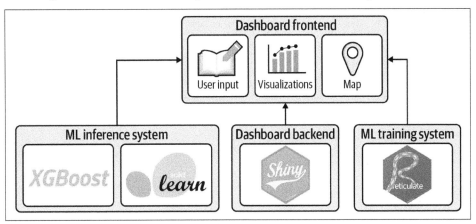

Figure 7-1. Architecture for our case study project

Alright, now that we know where we're going, let's choose our tools with care and assemble all the components into a unified whole.

We prepared this case study exclusively using the RStudio IDE. As we discussed in Chapter 6, if we're writing in R and accessing Python functions, this would be the way to go. The reason is the built-in capabilities in executing Python code chunks within R Markdown, the features of the Environment and Plot panes, and finally, the tooling around shiny.

Setup and Importing Data

We can see from our diagram that our end product will be an interactive R Markdown document. So let's begin as we did in Chapter 5. Our YAML header will consist of at least:[4]

```
---
title: "R & Python Case Study"
author: "Python & R for the modern data scientist"
runtime: shiny
---
```

 To have nicer formatting, we'll exclude the characters specifying an R Markdown chunk from the following examples. Naturally, if you are following along, you need to add them.

Since the data is stored in an SQLite database, we need to use some additional packages in addition to ones we've already seen. Our first code chunk is:

```
library(tidyverse)
library(RSQLite) # SQLite
library(DBI) # R Database Interface
```

In our second code chunk, we'll connect to the database and list all of the 33 available tables:

```
# Connect to an in-memory RSQLite database
con <- dbConnect(SQLite(), "ch07/data/FPA_FOD_20170508.sqlite")

# Show all tables
dbListTables(con)
```

Creating a connection (con) object is a standard practice in establishing programmatic access to databases. In contrast to R, Python has built-in support for opening such files with the sqlite3 package. This is preferable to R since we don't need to install and load two additional packages. Nonetheless, R is a core language for the initial steps, so we might as well just import the data in R from the outset.

Our data is stored in the Fires table. Because we know the columns we want to access, we can specify that while importing.

---

4 Some readers might not be familiar with this language. It is commonly used to specify configuration options as code, such as in this case.

It's also important to remember to close the connections when working with remote or shared databases because that might prevent other users from accessing the database and cause issues.[5]

```
fires <- dbGetQuery(con, "
                        SELECT
                        STAT_CAUSE_DESCR, OWNER_CODE, DISCOVERY_DOY,
                        FIRE_SIZE, LATITUDE, LONGITUDE
                        FROM Fires
                        WHERE (FIRE_YEAR=2015 AND STATE != 'AK' AND STATE !=
                            'HI');")
dbDisconnect(con)

dim(fires)
```

We limit our dataset size already at this very first importing step. It's a shame to throw out so much data. Still, we do this because older data, especially in climate applications, tends to be less representative of the current or near-future situation. Predictions based on old data can be inherently biased. By limiting the size of the dataset, we also reduce the amount of memory used, improving performance.

 Often in the case of enormous datasets (those barely or not fitting into the memory of your machine), you can use an ingestion command to select just a sample, such as LIMIT 1000.

We can get a preview of the data using the tidyverse function dplyr::glimpse():

```
glimpse(fires)

Rows: 73,688
Columns: 6
$ STAT_CAUSE_DESCR <chr> "Lightning", "Lightning", "Lightning", "Lightning"…
$ OWNER_CODE       <dbl> 5, 5, 5, 5, 5, 5, 5, 5, 5, 5, 5, 5, 5, 8, 5, 8, 5,…
$ DISCOVERY_DOY    <int> 226, 232, 195, 226, 272, 181, 146, 219, 191, 192, …
$ FIRE_SIZE        <dbl> 0.10, 6313.00, 0.25, 0.10, 0.10, 0.25, 0.10, 0.10,…
$ LATITUDE         <dbl> 45.93417, 45.51528, 45.72722, 45.45556, 44.41667, …
$ LONGITUDE        <dbl> -113.0208, -113.2453, -112.9439, -113.7497, -112.8…
```

---

5 This part can also be done very well within R by using packages such as dbplyr or the using the Connections panel in RStudio.

EDA and Data Visualization

Because the dataset is still relatively large, we should think carefully about the best data visualization strategy. Our first instinct may be to plot a map since we have latitude and longitude coordinates. This can be fed into ggplot2 directly as x- and y-axis coordinates as such:

```
g <- ggplot(fires, aes(x = LONGITUDE,
                       y = LATITUDE,
                       size = FIRE_SIZE,
                       color = factor(OWNER_CODE))) +
  geom_point(alpha = 0.15, shape = 16) +
  scale_size(range = c(0.5, 10)) +
  theme_classic() +
  theme(legend.position = "bottom",
        panel.background = element_rect(fill = "grey10"))

g
```

By mapping OWNER_CODE onto the color aesthetic (Figure 7-2), we can see a strong correlation in some states. We can predict that this will have a substantial effect on our model's performance.

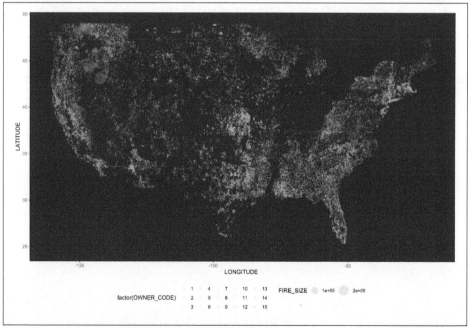

Figure 7-2. Plotting the sizes of individual fires

In the preceding code snippet, we assigned the plot to the object g. This is not strictly necessary, but we did it in this case to showcase the strength of the ggplot2 layering method. We can add a facet_wrap() layer to this plot and separate it into 13 facets, or *small multiples*, one for each type of STAT_CAUSE_DESCR (Figure 7-3):

```
g +
    facet_wrap(facets = vars(STAT_CAUSE_DESCR), nrow = 4)
```

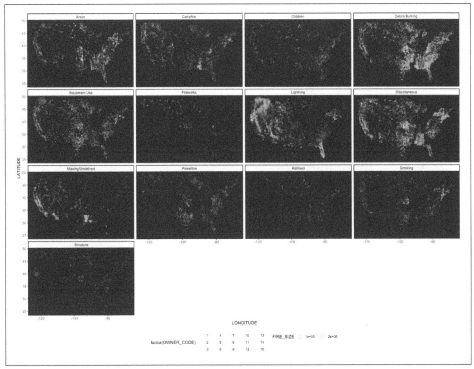

Figure 7-3. Faceting the fires plot, based on the fire cause

This allows us to appreciate that some causes are abundant while others are rare, an observation we'll see again shortly in a different way. We can also begin to assess any strong associations between, e.g., region, owner code, and cause of a fire.

Returning to the entirety of the dataset, an easy way to get a comprehensive overview is to use a pairs plot, sometimes called a *splom* (or scatter plot matrix if it consists of purely numeric data). The GGally package provides an exceptional function, ggpairs() that produces a matrix of plots (Figure 7-4).[6] Each pairwise bivariate plot

---

6 This package is used to extend the ggplot2 functionality for transformed datasets.

is shown as univariate density plots or histograms on the diagonal. In the upper triangle, the correlation between continuous features is available:

```
library(GGally)
fires %>%
  ggpairs()
```

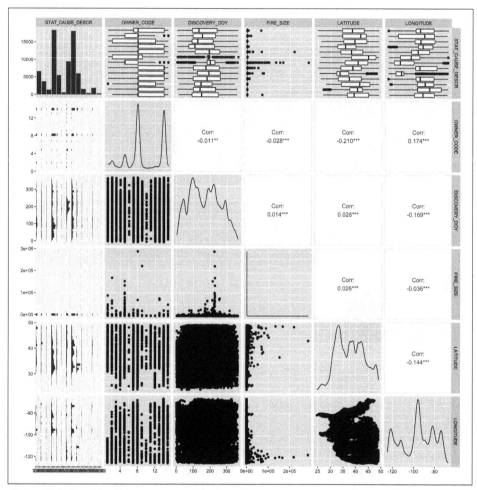

Figure 7-4. A pairs plot

This information-rich visualization demands some time to process. It's handy as an *exploratory* plot, in EDA, but not necessarily as an *explanatory* plot in reporting our results. Can you spot any unusual patterns? First, STAT_CAUSE_DESCR looks imbalanced, meaning there is a significant difference between the number of observations per class. Additionally, OWNER_CODE appears to be bimodal (having two maxima). Those properties can negatively affect our analysis, depending on which model we choose. Second, all correlations seem to be relatively low, making our job easier (since correlated data is not good for ML). Still, we already know there is a strong association between location (LATITUDE and LONGITUDE) and owner code from our previous plot. So we should take these correlations with a grain of salt. We would expect to detect this issue in feature engineering. Third, FIRE_SIZE has a very unusual distribution. It looks like that plot is empty, with just the x- and y-axes present. We see a density plot with a very high and narrow peak at the very low range and an extremely long positive skew. We can quickly generate a log_{10} transformed density plot (Figure 7-5):

```
ggplot(fires, aes(FIRE_SIZE)) +
  geom_density() +
  scale_x_log10()
```

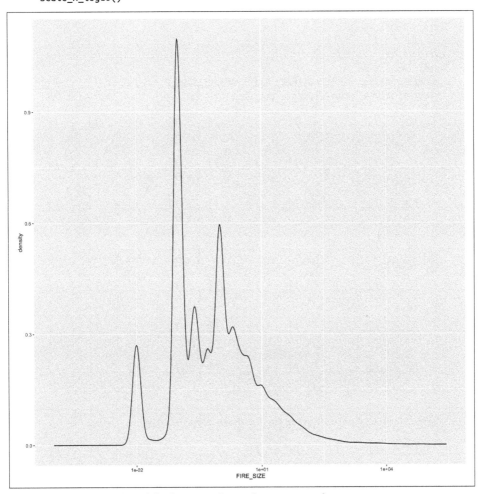

Figure 7-5. Density plot of the log-transformed FIRE_SIZE feature

For the case study, we'll keep the tasks to a minimum, but there might be a few other interesting things to visualize that can help tell a story for the end user. For example, note that the dataset has a temporal dimension. It would be interesting to see how forest fires' quantity (and quality) has been changing over time. We'll leave this to the motivated user to explore with the excellent gganimate package.

Interactive data visualization is often used without a special purpose in mind. Even for the most popular packages, the documentation shows just basic usage. In our case, since we have so many data points in a spatial setting, and we want to have a final deliverable that is accessible, creating an interactive map is an obvious choice. As in Chapter 5, we use Leaflet (Figure 7-6):

```
library(leaflet)

leaflet() %>%
  addTiles() %>%
  addMarkers(lng = df$LONGITUDE, lat = df$LATITUDE,
  clusterOptions = markerClusterOptions()
)
```

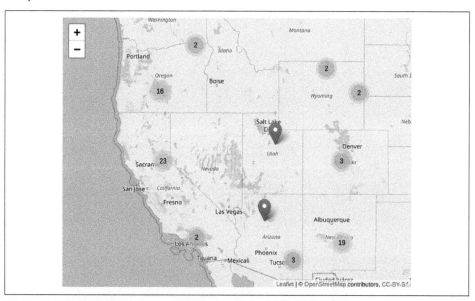

Figure 7-6. Interactive map showing the locations of forest fires[7]

---

7 A color version is available for print readers online (*https://oreil.ly/prmd_7-6*).

Note how using clusterOptions allows us to simultaneously present all of the data without overwhelming the user or reducing visibility. For our purposes, this satisfies our curiosity about using some great visualizations in EDA. There are plenty of other statistics we can apply, but let's move to machine learning in Python.

Machine Learning

By now, we have some idea about the factors that may influence the cause of a fire. Let's dive into building a machine learning model using scikit-learn in Python.[8]

We argued that ML is best done in Python as we saw in Chapter 5. We'll use a random forest algorithm. There are several reasons for this choice:

1. It's a well-established algorithm.
2. It's relatively easy to understand.
3. It does not require feature scaling before training.

There are other reasons why it's good, such as working well with missing data and having out-of-the-box explainability.

Setting Up Our Python Environment

As discussed in Chapter 6, there are a few ways to access Python using the reticulate package. The choice depends on the circumstances, which we laid out in our project architecture. Here, we'll pass our R data.frame to a Python virtual environment. If you followed the steps in Chapter 6, you'd already have the modern_data virtual environment set up. We already installed some packages into this environment. To recap, we executed the following commands:

```
library(reticulate)

# Create a new virtualenv
virtualenv_create("modern_data")

# Install Python packages into this virtualenv
library(tidyverse)
c("scikit-learn", "pandas", "seaborn") %>%
  purrr::map(~ virtualenv_install("modern_data", .))
```

---

8 This is not a thorough exposition of all possible methods or optimizations since our focus is on building a bilingual workflow, not exploring machine learning techniques in detail. Readers may choose to refer to the official scikit-learn documentation for further guidance, in the aptly named "Choosing the right estimator" (*https://oreil.ly/yVysu*).

If you don't have the `modern_data` virtualenv or you're using Windows, refer to the steps in the files *0—setup.R* and *1—activate.R* and discussed in Chapter 6. You may want to restart R at this point to make sure that you'll be able to activate your virtual environment using the following command:

```
# Activate virtual environment
use_virtualenv("modern_data", required = TRUE)

# If using miniconda (windows)
# use_condaenv("modern_data")
```

We'll include all the Python steps into a single script; you can find this script in the book's repository (*https://github.com/moderndatadesign/PyR4MDS*) under *ml.py*. First, we'll import the necessary modules:

```
from sklearn.ensemble import RandomForestClassifier
from sklearn.preprocessing import LabelEncoder
from sklearn.model_selection import train_test_split
from sklearn import metrics
```

Feature Engineering

There are features in the dataset that might be informative to a data analyst but are at best useless for training the model, and at worst can reduce its accuracy. This is called *adding noise* to the dataset, and we want to avoid it at all costs. This is the purpose behind feature engineering. Let's select just the features we need, as specified in Table 7-1. We also use standard ML conventions in storing them in X, and our target in y:

```
features = ["OWNER_CODE", "DISCOVERY_DOY", "FIRE_SIZE", "LATITUDE", "LONGITUDE"]
X = df[features]
y = df["STAT_CAUSE_DESCR"]
```

Here, we create an instance of the `LabelEncoder`. We use this to encode a categorical feature to numeric. In our case, we apply it to our target:

```
le = LabelEncoder()
y = le.fit_transform(y)
```

Here, we split the dataset into a training and a test set (note that we are also using the handy `stratify` parameter to make sure the splitting function samples our imbalanced classes fairly):

```
X_train, X_test, y_train, y_test = train_test_split(X, y, test_size=0.33,
                                                    random_state=42, stratify=y)
```

Model Training

To apply the random forest classifier, we'll make an instance of `RandomForestClassi`
`fier`. As in Chapter 5 we use the `fit/predict` paradigm and store the predicted values in `preds`:

```
clf = RandomForestClassifier()

clf.fit(X_train, y_train)

preds = clf.predict(X_test)
```

In the final step, we'll assign the confusion matrix and the accuracy score to objects:

```
conmat = metrics.confusion_matrix(y_test, preds)
acc = metrics.accuracy_score(y_test, preds)
```

After we have completed our script, we can source it into R:

```
source_python("ml.py")
```

After running this command, we'll have access to all the Python objects directly in
our environment. The accuracy is 0.58, which is not phenomenal, but certainly much
better than random!

When we use the `source_python` function from reticulate we can
significantly increase our productivity, especially if we are working
in a bilingual team. Imagine the scenario when a coworker of yours
builds the ML part in Python and you need to include their work in
yours. It would be as easy as sourcing without worrying about
recoding everything. This scenario is also plausible when joining a
new company or project and inheriting Python code that you need
to use straightaway.

If we want to take advantage of `ggplot` to examine the confusion matrix, we first need
to convert to an R `data.frame`. The `value` is then the number of observations of each
case, which we map onto `size`, and change the `shape` to 1 (a circle). The result is
shown in Figure 7-7:

```
library(ggplot2)
py$conmat %>%
  as.data.frame.table(responseName = "value") %>%
  ggplot(aes(Var1, Var2, size = value)) +
  geom_point(shape = 1)
```

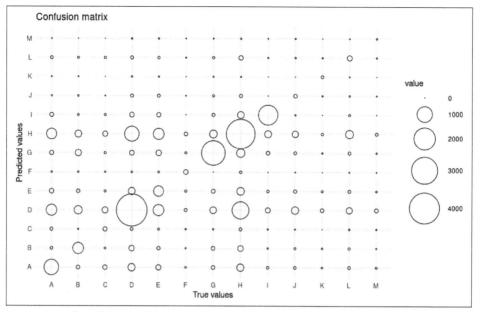

Figure 7-7. Plot of the classifier confusion matrix

It's not surprising that we have some groups with a very high match since we already knew that our data was imbalanced to begin with. Now, what do we do with this nice Python code and output? At the end of Chapter 6, we saw a simple and effective way to create an interactive document (remember what you learned in Chapter 5) using an R Markdown with a shiny runtime. Let's implement the same concept here.

Prediction and UI

Once we have established a Python model, it's general practice to test it with mock input. This allows us to ensure our model can handle the correct input data and is standard practice in ML engineering before connecting it with real user input. To this end, we'll create five sliderInputs for the five features of our model. Here, we've hardcoded the min and max values for the sake of simplicity, but these can, of course, be dynamic:

```
sliderInput("OWNER_CODE", "Owner code:",
        min = 1, max = 15, value = 1)
sliderInput("DISCOVERY_DOY", "Day of the year:",
        min = 1, max = 365, value = 36)
sliderInput("FIRE_SIZE", "Number of bins (log10):",
        min = -4, max = 6, value = 1)
sliderInput("LATITUDE", "Latitude:",
        min = 17.965571, max = 48.9992, value = 30)
sliderInput("LONGITUDE", "Longitude:",
        min = -124.6615, max = -65.321389, value = 30)
```

Similar to what we did at the end of Chapter 6, we'll access these values in the internal `input` list and use a shiny package function to render the appropriate output (Figure 7-8).

```
prediction <- renderText({
  input_df <- data.frame(OWNER_CODE = input$OWNER_CODE,
                         DISCOVERY_DOY = input$DISCOVERY_DOY,
                         FIRE_SIZE = input$FIRE_SIZE,
                         LATITUDE = input$LATITUDE,
                         LONGITUDE = input$LONGITUDE)

  clf$predict(r_to_py(input_df))
})
```

Those elements will respond dynamically to changes in user input. This is precisely what we need for our work since this is an interactive product and not a static one. You can see all of the different code blocks that we used in preparation for this project. They should require little change, with the most notable one being the ability to capture the user input in the inference part. This can be done by accessing the `input` object.

Final Thoughts

In this case study, we demonstrated how you could make the best of both worlds and combine the excellent tools that modern data scientists have at their disposal to create remarkable user experiences, which delight visually and inform decision making. This is but a basic example of such an elegant system, and we are confident that by showing you what's possible, you—our readers—will create the data science products of the future!

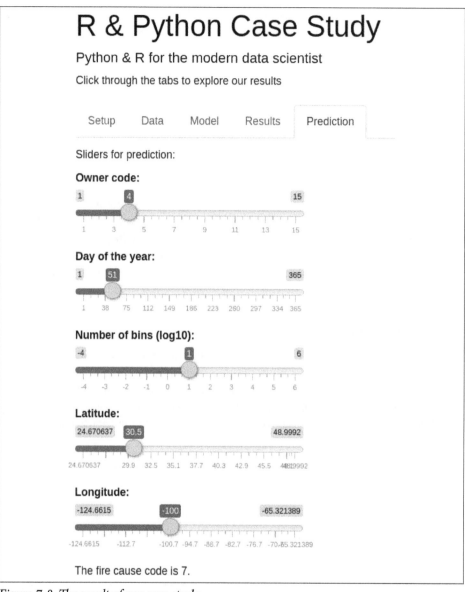

Figure 7-8. *The result of our case study*

A Python:R Bilingual Dictionary

The following dictionary is meant to be used as a quick reference for translating commands between Python and R. Visit the book's repo (*https://github.com/moderndatade sign/PyR4MDS*) for access to other resources.

Corrections and additions are welcome. Please contact Rick on LinkedIn (*https://www.linkedin.com/in/rick-scavetta*) or place an issue on the repo. For a downloadable summary, please visit the book's website (*https://moderndata.design*).

Aside from some command-line expressions to be entered in the terminal, which are explicitly noted, expressions are in R or Python.

Package Management

Table A-1. Installing a single package

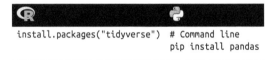

```
install.packages("tidyverse")  # Command line
                               pip install pandas
```

Table A-2. Installing specific package versions

```
devtools::install_version(    # Command line
  "ggmap",                     pip install pandas==1.1.0
  version = "3.5.2"
  )
```

Table A-3. Installing multiple packages

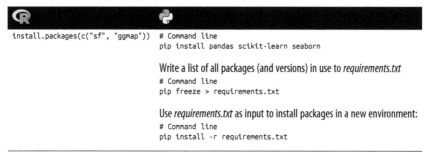

<image>	<image>

`install.packages(c("sf", "ggmap"))`	`# Command line`
	`pip install pandas scikit-learn seaborn`
	
	Write a list of all packages (and versions) in use to *requirements.txt*
	`# Command line`
	`pip freeze > requirements.txt`
	
	Use *requirements.txt* as input to install packages in a new environment:
	`# Command line`
	`pip install -r requirements.txt`

Table A-4. Loading packages

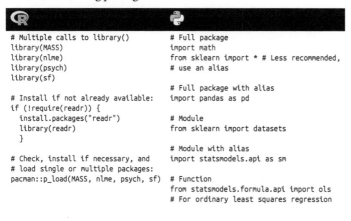

<image>	<image>

`# Multiple calls to library()`
`library(MASS)`
`library(nlme)`
`library(psych)`
`library(sf)`
 	
`# Install if not already available:`
`if (!require(readr)) {`
`  install.packages("readr")`
`  library(readr)`
`  }`
 	
`# Check, install if necessary, and`
`# load single or multiple packages:`
`pacman::p_load(MASS, nlme, psych, sf)`
	`# Full package`
	`import math`
	`from sklearn import * # Less recommended,`
	`# use an alias`
	
	`# Full package with alias`
	`import pandas as pd`
	
	`# Module`
	`from sklearn import datasets`
	
	`# Module with alias`
	`import statsmodels.api as sm`
	
	`# Function`
	`from statsmodels.formula.api import ols`
	`# For ordinary least squares regression`

Assign Operators

Table A-5. Typical assign operators in R[a]

Operator	Direction	Environment	Name	Comment
`<-`	RHS to LHS	Current	Assignment operator (leftwards)	Preferred; common and unambiguous
`=`	RHS to LHS	Current	Assignment operator (leftwards)	Less preferred; common but easily confused with == (equivalency) and = (assign to function argument); no corollary super assignment
`->`	LHS to RHS	Current	Assignment operator (rightwards)	Less preferred; uncommon, easily overlooked, and unexpected. Often used at the end of a long dplyr/tidyverse chain of functions; choose %<% instead.

[a] RHS stands for right-hand side and LHS stands for left-hand side.

Table A-6. Typical assign operator in Python

Operator	Direction	Environment	Name	Comment
=	RHS to LHS	Current	Simple assignment operator	Preferred; use following environment scoping rules

Table A-7. Super assignment operators in R

Operator	Direction	Environment	Name	Comment
<<-	RHS to LHS	Parent	Super assignment operator (leftwards)	Common; use following environment scoping rules
->>	LHS to RHS	Parent	Super assignment operator (rightwards)	Less common

These operators are particularly preferred when using a dplyr/tidyverse chain of functions.

Table A-8. Special cases in R

Operator	Direction	Environment	Name	Comment
|>	LHS to RHS	Current	Native forward pipe	Assign to the first argument of the downstream function
%>%	LHS to RHS	Current	Forward pipe or pipe (colloquially)	Assign to the first argument of the downstream function, magrittr package
%$%	LHS to RHS	Current	Exposition pipe	Expose the named elements to the downstream function, magrittr package
%<>%	RHS to LHS	Current	Assignment pipe	Assign to the first argument of the downstream function and assign output in situ, magrittr package
%<-%	RHS to LHS	Current	Multiple assign	Assign to multiple objects, zeallot package

Table A-9. Special cases and incrementals in Python

Operator	Direction	Environment	Name	Comment
+=	RHS to LHS	Current	Increment assignment	Adds a value and the variable and assigns the result to that variable.
-=	RHS to LHS	Current	Decrement assignment	Subtracts a value from the variable and assigns the result to that variable.
*=	RHS to LHS	Current	Multiplication assignment	Multiplies the variable by a value and assigns the result to that variable.
/=	RHS to LHS	Current	Division assignment	Divides the variable by a value and assigns the result to that variable.
**=	RHS to LHS	Current	Power assignment	Raises the variable to a specified power and assigns the result to that variable.

Operator	Direction	Environment	Name	Comment
%=	RHS to LHS	Current	Modulus assignment	Computes the modulus of the variable and a value and assigns the result to that variable.
//=	RHS to LHS	Current	Floor division assignment	Floor divides the variable by a value and assigns the result to that variable.

Types

Table A-10. *The four most common user-defined atomic-vector types in R*

Type	Data frame shorthand	Tibble shorthand	Description	Example
Logical	logi	<lgl>	Binary data	TRUE/FALSE, T/F, 1/0
Integer	int	<int>	Whole numbers from $-\infty, \infty$	7, 9, 2, −4
Double	num	<dbl>	Real numbers from $-\infty, \infty$	3.14, 2.78, 6.45
Character	chr	<chr>	All alpha-numeric characters, including white spaces	"Apple," "Dog"

Table A-11. *The four most common user-defined types in Python*

Type	Shorthand	Description	Example
Boolean	bool	Binary data	True/False
Integer	int	Whole numbers from $-\infty, \infty$	7, 9, 2, −4
Float	float	Real numbers from $-\infty, \infty$	3.14, 2.78, 6.45
String	str	All alpha-numeric characters, including white spaces	"Apple," "Dog"

Arithmetic Operators

Table A-12. *Common arithmetic operators*

Description	R Operator	Python Operator
Addition	+	+
Subtraction	−	−
Multiplication	*	*
Division (float)	/	/
Exponentiation	^ or **	**
Integer Division (floor)	%/%	//
Modulus	%%	%

Attributes

Table A-13. Class attributes

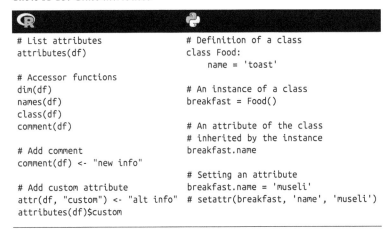

```
# List attributes                    # Definition of a class
attributes(df)                       class Food:
                                         name = 'toast'
# Accessor functions
dim(df)                              # An instance of a class
names(df)                            breakfast = Food()
class(df)
comment(df)                          # An attribute of the class
                                     # inherited by the instance
# Add comment                        breakfast.name
comment(df) <- "new info"
                                     # Setting an attribute
# Add custom attribute               breakfast.name = 'museli'
attr(df, "custom") <- "alt info"     # setattr(breakfast, 'name', 'museli')
attributes(df)$custom
```

Keywords

Table A-14. Reserved words and keywords

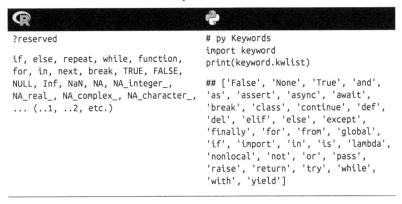

```
?reserved                            # py Keywords
                                     import keyword
if, else, repeat, while, function,   print(keyword.kwlist)
for, in, next, break, TRUE, FALSE,
NULL, Inf, NaN, NA, NA_integer_,     ## ['False', 'None', 'True', 'and',
NA_real_, NA_complex_, NA_character_, 'as', 'assert', 'async', 'await',
... (..1, ..2, etc.)                 'break', 'class', 'continue', 'def',
                                     'del', 'elif', 'else', 'except',
                                     'finally', 'for', 'from', 'global',
                                     'if', 'import', 'in', 'is', 'lambda',
                                     'nonlocal', 'not', 'or', 'pass',
                                     'raise', 'return', 'try', 'while',
                                     'with', 'yield']
```

Functions and Methods

Table A-15. Defining named functions

R	Python
| ```r
# Basic definition
myFunc <- function (x, ...) {
  x * 10
}

myFunc(4)

## [1] 40

# Multiple unnamed arguments
myFunc <- function (...) {
  sum(...)
}

myFunc(100,40,60)

## [1] 200
``` | ```python
# Simple definition
def my_func(x):
    return(x * 10)

my_func(4)

## 40

# Multiple named arguments, passed as a tuple
def my_func(*x):
    return(x[2])

my_func(100, 40, 60)

## 60

# Multiple unknown arguments, saved as a dict
def my_func(**numb):
    print("x: ",numb["x"])
    print("y: ",numb["y"])

my_func(x = 40, y = 100)

## x:   40
## y:   100

# Using doc strings
def my_func(**numb):
    """An example function
    that takes multiple unknown arguments.
    """
    print("x: ",numb["x"])
    print("y: ",numb["y"])

# Access doc strings with dunder
my_func.__doc__

'An example function
    that takes multiple unknown arguments.'
``` |

Style and Naming Conventions

Style in R is generally more loosely defined than in Python. Nonetheless, see the *Advanced R style guide* (*https://oreil.ly/BOf3M*) by Hadley Wickham (CRC Press) or Google's R Style guide (*https://oreil.ly/ZyXjb*) for suggestions.

For Python, see the PEP 8 style guide (*https://oreil.ly/UqWqs*).

Table A-16. Style and naming conventions in R and Python

Indentation and spacing	Naming in a script	Indentation and spacing	Naming in a script
White space is generally for style and inconsequential to execution. Add a space around operators and use a tab to indent on successive lines of long commands.	The trend is currently toward lowercase snake case: underscores ("_") between words and only lowercase letters. Example: `my_data <- 1:6`	White space, in particular indentation, is a part of Python execution. Use four spaces instead of a tab (this can be set in your text editor).	Type: Functions and variables Style: Lowercase snake case Example: `func`, `my_func`, `var`, `my_var`, `x`

Table A-17. When defining classes

Type	Style	Example
Class	Capitalized camel case	`Recipe`, `MyClass`
Method	Lowercase snake case	`class_method`, `method`
Constant	Full uppercase snake case	`CONS`, `MY_CONS`, `LONG_NAME_CONSTANT`

Table A-18. In packages

Type	Style	Example
Packages and modules	Lowercase snake case	`mypackage`, `module.py`, `my_module.py`

Table A-19. Naming conventions with _

Naming	Meaning
`_var`	A convention used to show that a variable is meant for internal use within a function or method
`var_`	A convention used to avoid naming conflicts with Python keywords
`__var`	Triggers name mangling when used in a class context to prevent inheritance collisions. Enforced by the Python interpreter
`__var__`	*Dunder* ("double underscore") variables. Special methods defined by the Python language. Avoid this naming scheme for your own attributes.
`_`	Naming a temporary or insignificant variable, e.g., in a for loop

Analogous Data Storage Objects

Table A-20. Analogous Python objects for common R objects

®Structure	🐍 Analogous structure(s)
Vector (one-dimensional homogeneous)	ndarray, but also scalars, homogeneous list and tuple
Vector, matrix or array (homogeneous)	NumPy n-dimensional array (ndarray)
Unnamed list (heterogenous)	list
Named list (heterogeneous)	Dictionary dict, but lacking order
Environment (named, but unordered elements)	Dictionary dict
Variable/column in a data.frame	Pandas Series (pd.Series)
Two-dimensional data.frame	Pandas data frame (pd.DataFrame)

Table A-21. Analogous R objects for common Python objects

🐍 Structure	® Analogous structure(s)
scalar	One-element long vector
list (homogeneous)	Vector, but as if lacking vectorization
list (heterogeneous)	Unnamed list
tuple (immutable, homogeneous)	Vector, list as separated output from a function
Dictionary dict, a key-value pair	Named list or better environment
NumPy n-dimensional array (ndarray)	Vector, matrix, or array
Pandas Series (pd.Series)	Vector, variable/column in a data.frame
Pandas data frame (pd.DataFrame)	Two-dimensional data.frame

Table A-22. One-dimensional, homogeneous

®	🐍
| ```
# Vectors
cities_R <- c("Munich", "Paris", "Amsterdam")
dist_R <- c(584, 1054, 653)
``` | ```
# Lists
cities = ['Munich', 'Paris', 'Amsterdam']
dist = [584, 1054, 653]
``` |

Table A-23. One-dimensional, heterogeneous key-value pairs (Lists in R, dictionaries in Python)

®	🐍

<pre># A list of data frames 	
cities_list <- list(Munich = data.frame(dist = 584,
                                       pop = 1484226,
                                       area = 310.43,
                                       country = "DE"),
                     Paris = data.frame(dist = 1054,
                                       pop = 2175601,
                                       area = 105.4,
                                       country = "FR"),
                     Amsterdam = data.frame(dist = 653,
                                       pop = 1558755,
                                       area = 219.32,
                                       country = "NL"))
 	
# As a list object 	
cities_list[1]
 	
## $Munich 	
## dist pop area country 	
## 1 584 1484226 310.43 DE 	
 	
cities_list["Munich"]
 	
## $Munich 	
## dist pop area country 	
## 1 584 1484226 310.43 DE 	
 	
# As a data.frame object 	
cities_list[[1]]
 	
## dist pop area country 	
## 1 584 1484226 310.43 DE 	
 	
cities_list$Munich
 	
## dist pop area country 	
## 1 584 1484226 310.43 DE 	
 	
# A list of heterogeneous data 	
lm_list <- lm(weight ~ group, data = PlantGrowth)
 	
# length(lm_list) 	
# names(lm_list)</pre> 	
	<pre># lists
	city_l = ['Munich', 'Paris', 'Amsterdam']
	
	dist_l = [584, 1054, 653]
	
	pop_l = [1484226, 2175601, 1558755]
	
	area_l = [310.43, 105.4, 219.32]
	
	country_l = ['DE', 'FR', 'NL']
	
	import numpy as np
	
	# NumPy arrays
	city_a = np.array(['Munich', 'Paris', 'Amsterdam'])
	city_a
	
	## array(['Munich', 'Paris', 'Amsterdam'], dtype=
	## '<U9')
	
	pop_a = np.array([1484226, 2175601, 1558755])
	pop_a
	
	## array([1484226, 2175601, 1558755])
	
	# Dictionaries
	yy = {'city': ['Munich', 'Paris', 'Amsterdam'],
	      'dist': [584, 1054, 653],
	      'pop': [1484226, 2175601, 1558755],
	      'area': [310.43, 105.4, 219.32],
	      'country': ['DE', 'FR', 'NL']}
	yy
	
	## {'city': ['Munich', 'Paris', 'Amsterdam'],
	## 'dist': [584, 1054, 653], 'pop': [1484226,
	## 2175601, 1558755], 'area': [310.43, 105.4,
	## 219.32], 'country': ['DE', 'FR', 'NL']}</pre>

Data Frames

Table A-24. Data frames in Python

```
# class pd.DataFrame
import pandas as pd

# From a dictionary, yy
yy_df = pd.DataFrame(yy)
yy_df

##           city  dist       pop    area country
## 0       Munich   584  1484226  310.43      DE
## 1        Paris  1054  2175601  105.40      FR
## 2    Amsterdam   653  1558755  219.32      NL

# From lists
# names
list_names = ['city', 'dist', 'pop', 'area', 'country']

# columns are a list of lists
list_cols = [city_l, dist_l, pop_l, area_l, country_l]
list_cols


## [['Munich', 'Paris', 'Amsterdam'], [584, 1054, 653], [1484226...

# A zipped list of tuples

zip_list = list(zip(list_cols, list_names))
zip_list

# zip_dict = dict(zip_list)
# zip_df = pd.DataFrame(zip_dict)
# zip_df


# zip_df = pd.DataFrame(zip_list)
# zip_df


## [(['Munich', 'Paris', 'Amsterdam'], 'city'), ([584, 1054, 653], 'dist')...
```

```python
# Easier
# Import pandas library
import pandas as pd

# initialize list of lists
list_rows = [['Munich',    584, 1484226, 310.43, 'DE'],
             ['Paris',    1054, 2175601, 105.40,      'FR'],
             ['Amsterdam',  653, 1558755, 219.32,      'NL']]

# Create the pandas data frame
df = pd.DataFrame(list_rows, columns = list_names)

# print data frame.
df


##          city  dist       pop    area country
## 0      Munich   584  1484226  310.43      DE
## 1       Paris  1054  2175601  105.40      FR
## 2   Amsterdam   653  1558755  219.32      NL
```

Table A-25. Two-dimensional, heterogenous, tabular data frames in R

```r
# class data.frame from vectors
cities_df <- data.frame(city = c("Munich", "Paris", "Amsterdam"),
                        dist = c(584, 1054, 653),
                        pop = c(1484226, 2175601, 1558755),
                        area = c(310.43, 105.4, 219.32),
                        country = c("DE", "FR", "NL"))

cities_df

##        city dist      pop   area country
## 1    Munich  584 1484226 310.43      DE
## 2     Paris 1054 2175601 105.40      FR
## 3 Amsterdam  653 1558755 219.32      NL
```

Table A-26. Multidimensional arrays

R	Python

```r
# array
arr_r <- array(c(1:4,
                 seq(10, 40, 10),
                 seq(100, 400, 100)),
               dim = c(2,2,3) )

arr_r

## , , 1
##
##      [,1] [,2]
## [1,]    1    3
## [2,]    2    4
##
## , , 2
##
##      [,1] [,2]
## [1,]   10   30
## [2,]   20   40
##
## , , 3
##
##      [,1] [,2]
## [1,]  100  300
## [2,]  200  400

rowSums(arr_r, dims = 2)

##      [,1] [,2]
## [1,]  111  333
## [2,]  222  444

rowSums(arr_r, dims = 1)

## [1] 444 666

colSums(arr_r, dims = 1)

##      [,1] [,2] [,3]
## [1,]    3   30  300
## [2,]    7   70  700

colSums(arr_r, dims = 2)

## [1]   10  100 1000
```

```python
arr = np.array([[[  1,   2],
                 [  3,   4]],
                [[ 10,  20],
                 [ 30,  40]],
                [[100, 200],
                 [300, 400]]])

arr

## array([[[  1,   2],
##         [  3,   4]],
##
##        [[ 10,  20],
##         [ 30,  40]],
##
##        [[100, 200],
##         [300, 400]]])

arr.sum(axis=0)

## array([[111, 222],
##        [333, 444]])

arr.sum(axis=1)

## array([[  4,   6],
##        [ 40,  60],
##        [400, 600]])

arr.sum(axis=2)

## array([[  3,   7],
##        [ 30,  70],
##        [300, 700]])
```

Logical Expressions

Table A-27. Relational operators

Description	R Operator	Python Operator
Equivalency	==	==
Non-equivalency	!=	!=
Greater-than (or equal to)	> (>=)	> (>=)
Lesser-than (or equal to)	< (<=)	< (<=)
Negation	!x	not()

Table A-28. Relational operators

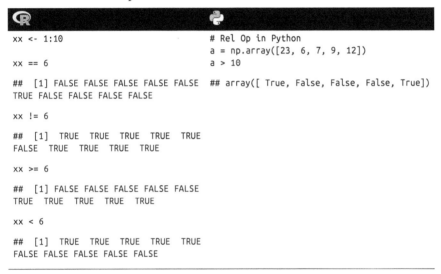

R	Python

`xx <- 1:10`
 	
`xx == 6`
 	
`## [1] FALSE FALSE FALSE FALSE FALSE`
`TRUE FALSE FALSE FALSE FALSE`
 	
`xx != 6`
 	
`## [1] TRUE TRUE TRUE TRUE TRUE`
`FALSE TRUE TRUE TRUE TRUE`
 	
`xx >= 6`
 	
`## [1] FALSE FALSE FALSE FALSE FALSE`
`TRUE TRUE TRUE TRUE TRUE`
 	
`xx < 6`
 	
`## [1] TRUE TRUE TRUE TRUE TRUE`
`FALSE FALSE FALSE FALSE FALSE`
	`# Rel Op in Python`
	`a = np.array([23, 6, 7, 9, 12])`
	`a > 10`
	
	`## array([ True, False, False, False, True])`

Table A-29. Logical operators

Description	R operator	Python operator
AND	&, &&	&, and
OR	|, ||	|, or
WITHIN	y %in% x	in, not in
identity	identical()	is, is not

Table A-30. Logical operators

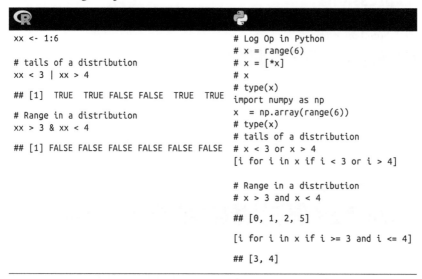

<image>	<image>
| ```
xx <- 1:6

# tails of a distribution
xx < 3 | xx > 4

## [1]  TRUE  TRUE FALSE FALSE  TRUE  TRUE

# Range in a distribution
xx > 3 & xx < 4

## [1] FALSE FALSE FALSE FALSE FALSE FALSE
``` | ```
# Log Op in Python
# x = range(6)
# x = [*x]
# x
# type(x)
import numpy as np
x  = np.array(range(6))
# type(x)
# tails of a distribution
# x < 3 or x > 4
[i for i in x if i < 3 or i > 4]

# Range in a distribution
# x > 3 and x < 4

## [0, 1, 2, 5]

[i for i in x if i >= 3 and i <= 4]

## [3, 4]
``` |

Table A-31. Identity

<image>	<image>
| ```
x <- c("Caracas", "Bogotá", "Quito")
y <- c("Bern", "Berlin", "Brussels")
z <- c("Caracas", "Bogotá", "Quito")

# Are the objects identical?
identical(x, y)

## [1] FALSE

identical(x, z)

## [1] TRUE

# Is any TRUE
any(x == "Quito")

## [1] TRUE

# Are all TRUE
all(str_detect(y, "^B"))

## [1] TRUE
``` | ```
x = ['Caracas', 'Bogotá', 'Quito']
y = ['Bern', 'Berlin', 'Brussels']
z = ['Caracas', 'Bogotá', 'Quito']

x == y

## False

x == z

## True

# Is any True
import numpy as np
x = np.array(x)
np.any(x == "Caracas")

## True

# Are all True

np.all(x == "Caracas")

## False
``` |

Indexing

Table A-32. Testing for identical objects[a]

R			Python		
Dimensions	Use	Description	Dimensions	Use	Description
1	x[index]	Isolate contents, keep container	1	x[index]	Isolate contents, keep container
1	x[-index]	Extract one content, discard container	1	x[-index]	Isolate contents from reverse direction, keep container
1	x[[index]]	Isolate contents, remove item, keep container	1	x[index_1:index_2]	Slice
2	x[row_index, col_index]	Isolate contents, keep container	1	x[index_1:index_2:stride]	Slice with an interval
2	x[col_index]	Shortcut for columns	1	x[index_1:index_2:-1]	Slice with reversal
2	x[[index]]	Extract one content, discard container	2	x.loc[index_1:index_2]	Location
n	x[row_index, col_index, dim_index]	Isolate contents, keep container	2	x.iloc[index_1:index_2:stride]	Index

[a] Where index, row_index, col_index, and dim_index are vectors of type integer, character, or logical

Table A-33. One-dimensional

R	Python
```xx <- LETTERS[6:16]``` ```xx[4]```	```cities = ['Toronto', 'Santiago',``` ```'Berlin', 'Singapore', 'Kampala', 'New Delhi']```
```## [1] "I"```	```cities[0]```
```xx[[4]]```	```## 'Toronto'```
```## [1] "I"```	```cities[-1]```
```cities_list[2]```	```## 'New Delhi'```

```## $Paris```
```##   dist   pop area country```
```## 1 1054 2175601 105.4   FR```
	```cities[1:2]```

	```## ['Santiago']```
	```cities[:2]```
	```## ['Toronto', 'Santiago']```

Table A-34. Two-dimensional R

```
R
```

```
# class data.frame from vectors
cities_df <- data.frame(city = c("Munich", "Paris", "Amsterdam"),
                    dist = c(584, 1054, 653),
                    pop = c(1484226, 2175601, 1558755),
                    area = c(310.43, 105.4, 219.32),
                    country = c("DE", "FR", "NL"))

cities_df[2] # data frame

##   dist
## 1  584
## 2 1054
## 3  653

cities_df[,2]  # vector

## [1]  584 1054  653

cities_df[[2]] # vector

## [1]  584 1054  653

cities_df[2:3] # data frame

##   dist     pop
## 1  584 1484226
## 2 1054 2175601
## 3  653 1558755

cities_df[,2:3]  # data frame

##   dist     pop
## 1  584 1484226
## 2 1054 2175601
## 3  653 1558755
```

```r
cities_tbl <- tibble(city = c("Munich", "Paris", "Amsterdam"),
                     dist = c(584, 1054, 653),
                     pop = c(1484226, 2175601, 1558755),
                     area = c(310.43, 105.4, 219.32),
                     country = c("DE", "FR", "NL"))


cities_tbl[2]   # data frame

## # A tibble: 3 x 1
##     dist
##    <dbl>
## 1    584
## 2   1054
## 3    653

cities_tbl[,2]   # data frame

## # A tibble: 3 x 1
##     dist
##    <dbl>
## 1    584
## 2   1054
## 3    653

cities_tbl[[2]] # vector

## [1]   584 1054   653

cities_tbl[2:3]   # data frame

## # A tibble: 3 x 2
##     dist     pop
##    <dbl>   <dbl>
## 1    584 1484226
## 2   1054 2175601
## 3    653 1558755

cities_tbl[,2:3]   # data frame

## # A tibble: 3 x 2
##     dist     pop
##    <dbl>   <dbl>
## 1    584 1484226
## 2   1054 2175601
## 3    653 1558755
```

Table A-35. Two-dimensional Python

```
df

##         city  dist      pop   area country
   ## 0    Munich   584  1484226  310.43      DE
   ## 1     Paris  1054  2175601  105.40      FR
   ## 2 Amsterdam   653  1558755  219.32      NL

df[1:]

##         city  dist      pop   area country
   ## 1     Paris  1054  2175601  105.40      FR
   ## 2 Amsterdam   653  1558755  219.32      NL

# position
  df.iloc[0, 1]

## 584

df.iat[0, 1]

## 584

# label
  df.loc[1:,  'city']

## 1         Paris
   ## 2     Amsterdam
   ## Name: city, dtype: object

data = {'Country': ['Belgium', 'India', 'Brazil'],
        'Capital': ['Brussels', 'New Delhi', 'Brasilia'],
        'Population': [11190846, 1303171035, 207847528]}

df_2 = pd.DataFrame(data,columns=['Country', 'Capital', 'Population'])

df_2

##     Country    Capital  Population
## 0  Belgium   Brussels    11190846
## 1    India  New Delhi  1303171035
## 2   Brazil   Brasilia   207847528

df[1:]
# df.iloc([0], [0])

##         city  dist      pop   area country
## 1     Paris  1054  2175601  105.40      FR
## 2 Amsterdam   653  1558755  219.32      NL
```

Table A-36. N-dimensional

| ```
cities_array <- c(1:16)
dim(cities_array) <- c(4,2,2)
cities_array

## , , 1
##
##      [,1] [,2]
## [1,]    1    5
## [2,]    2    6
## [3,]    3    7
## [4,]    4    8
##
## , , 2
##
##      [,1] [,2]
## [1,]    9   13
## [2,]   10   14
## [3,]   11   15
## [4,]   12   16

cities_array[1,2,2]

## [1] 13

cities_array[1,2,]

## [1]  5 13

cities_array[,2,1]

## [1] 5 6 7 8
``` | ```
# Python n-dimensional indexing
arr

## array([[[  1,    2],
##         [  3,    4]],
##
##        [[ 10,   20],
##         [ 30,   40]],
##
##        [[100, 200],
##         [300, 400]]])

arr[1,1,1]

## 40

arr[:,1,1]

## array([  4,   40, 400])

arr[1,:,1]

## array([20, 40])

arr[1,1,:]

## array([30, 40])
``` |

Index

arrays, Python NumPy
 broadcasting with, 32
 conversion to R matrix, 126
 image measuring and manipulation with,
  80-81
 masking in, 35
 OpenCV's storage of images as, 79
 quick reference to, 157
 vectorization in, 65
arrays, R, 24, 157
"as" keyword, Python, 59
as.* functions, R, 25
assign operators, 18, 65, 148
atomic vector types, R, 24-26
attach function, R, 20
attributes, R and Python, 150
Australian daily temperatures dataset, 77, 86-89
AWS Elastic Beanstalk, 109

B
base R
 old functions in, 19, 39-40
 R Core Team and, 4
 versus Tidyverse, 8, 19, 21-24, 27-28, 30, 36,
  40, 74, 76
 time series plotting in, 86-88
Bell Laboratories, 3
BentoML, 106
bilingualism in Python and R
 benefits of, ix-xi, 9, 89, 92, 143, 145
 case study of, 131
  (see also case study of R and Python
   interoperability)
 fundamental elements of, 11
 interoperability of, 117
  (see also interoperability between
   Python and R)
 versus language wars, 6-9
 modern context for, 69
  (see also data formats; workflows)
 process of becoming, xi-xiii, 11, 13, 43
 quick reference to, 147-165
bioclimatic data, 90
BioConductor, R, 6, 8, 79
bioinformatics, 6, 8
boolean (bool) data type, Python, 25, 35, 64,
 150
box plots, Python, 66
broadcasting, Python, 32

C
case study of R and Python interoperability,
 131-145
 overview of steps and architecture for, 133
 Python for featuring engineering in, 142
 Python for machine learning in, 141-144
 R for data import in, 134-136
 R for EDA and data visualization in,
  136-141
 R for interactive web interface for, 144-145
 R for mapping in, 140
 RStudio for, 133
 US Wildfires dataset for (see US Wildfires
  dataset)
categorical variables, 31
Chambers, John, 3, 5
character data type, R, 25, 31, 150
character vectors, R, 33, 35
class of type integer, R factor as, 31
class of type list, R data.frame as, 28, 31
classes of objects, R, 23
classification task, R, 90
Cleveland, William, 4
clf.predict method, Python, 127
cloud-based tools
 options for, 14, 46, 57, 109
 with Python, 46, 57, 109
 with R, 14
coding
 and avoiding hard-coding, 119
 in Python, 46, 55, 57
 from scratch, advantages, 76
colon (:), R, 34
columns in data frames, 27-28, 35, 60-62
command line coding, Python, 46, 55
Command Palette in text editors, 49
commands, executing in R, 14, 118
communication
 challenges of data scientists with, 93, 109
 with reports (see reporting results)
communities of users, Python and R, x, xii, 3, 5,
 9, 17
comparisons, statistical, 32, 67
Computational Methods for Data Analysis
 (Chambers), 3
computer vision (CV), 77
conda, Python and R, 45, 51, 55, 123
confusion matrix, 143
connection (con) object, 134

convolutional neural networks (CNNs), 73
correlations between variables, 137-139
CRAN (Comprehensive R Archive Network), 4,
    19, 55, 81, 121
CRISP-DM framework, 94
cropping (cutting) data, 90
CSV files
    importing and exporting, 120
    with module and read function, Python, 75
    with read functions, R, 15, 18, 22, 23
CV (computer vision), 77

D
dashboards, reporting with, 109
data analysis, programming languages for, 4
data augmentation, 78
data coercion, 83, 86
data engineering (DE)
    overview of packages for, 94
    Python tools as best with, 105-109
    workflows of, 105-109
data formats, 71-92
    accessible storage of, 79, 83
    best tools of Python and R for, 91
    defined, 72
    external versus base packages for, 74-77, 79
    importance of decisions about, 72-74
    outputs during processing of, 84
    overview of packages for, 74
    overview of use cases and, 76
    and package design, 92
    packages for image, 74, 77-82
    packages for spatial, 74, 89-91
    packages for text, 74, 82-86
    packages for time series, 74, 86-89
    pipelines for, 73
    selection criteria for, 71
    simplicity during processing of, 81, 84
    tabular, 21, 28, 74, 75, 97, 99
data frame indexing , Python pandas, 86
data frames
    Python (see pd.DataFrame objects, Python)
    quick reference to, 157
    R (see data.frame objects, R)
data lineage, tracking, 79, 132
data munging, 77, 87, 94
data pipelines, 73
data processing, importance of data format in,
    72-74

Data Science from Scratch (Grus), 76
data science milestones, timeline of, 9
data science, modern, ix-xi
data scientists
    communication challenges of, 93, 109
    community-building among, 9
    and data engineers, 105
    increasing productivity of, ix-xi, 74, 81, 92,
        102, 143
    and machine learning (ML), 100
    reports, options for, 109-113
    and using GUIs, 95
data storage objects (see objects)
data structures
    influence of, 7
    in Python, 62-65
    in R, 24, 28-31
data types
    defined, 72
    heterogeneous, 24, 30, 155
    homogeneous, 24, 154
    in Python, 64, 150
    in R, 21, 24-26, 150
data visualizations (DV), 4
    (see also plotting)
    in case study of R and Python interoperabil-
        ity, 136-141
    commercial versus open source tools for, 95
    with ggplot2 (see ggplot2 package, R)
    with GUIs, 17, 95
    interactive, 99-100, 144-145
    for ML model performance, 102-104,
        143-145
    need for, 95
    static, 96-98
data.frame function, R, 26, 27
data.frame objects, R
    dplyr and, 40
    example, 24
    indexing and, 35-36
    lists and, 28-31
    naming columns of, 27-28
    Python DataFrame, similarities to, 23, 62
    Python, conversion to, 120, 126, 127, 141
    quick reference to, 157
    rows and columns in, 35
    transformation to time series data, 86
    views/output of, 16, 17, 22, 29-31, 143
data.table package, R, 76

Python, 60, 75, 78, 119, 134
R, 27, 134-136
importing packages
    in Python, 20, 58-60, 148
    in R, 20, 23, 148
    with specific functions within packages, 23, 60
impostor syndrome, 93
index location (.iloc) method, Python, 65
indexing
    with 0 as beginning, Python, 61, 65
    with 1 as beginning, R, 23, 29
    purpose of, 11
    in Python, 62, 65, 160-165
    quick reference to, 160-165
    in R, 33-36, 37, 160-165
inferential statistics, 67
info method, Python, 61
initializing (loading) packages (see importing packages)
install.packages function, R, 19, 55, 56, 147
installations
    in Python, 47-49, 55-57, 147-148
    in R, 14-15, 19-21, 55, 147-148
integer (int) data type, Python, 64, 150
integer data type, R, 24-26, 31, 150
integer vectors, R, 33-35
integrated development environments (IDEs) (see IDEs (integrated development environments))
interactive modes
    for calling Python from R, 128-130
    with data visualizations, 99-100, 140, 144-145
    with reporting results, 111-113
    with web interface, 144-145
interoperability between Python and R, 117-130
    by calling Python, 127-130
    by passing objects, 125
    with predefined scripts, 117-120
    problems to avoid with, 119, 128
    with Python REPL console, 128
    with R Markdown document, 125-127
    with reticulate package for R to Python, 120-125, 141
    with rpy2 package for Python to R, 120
    with shiny interactivity, 129, 133, 144-145
    by sourcing Python scripts, 127, 143

virtual environment for, 122-124
interoperability within Python ecosystem, 79, 82, 102, 104, 109
interpreter, executing commands in, 128
ipyKernel in VS Code, 53
IPython package, 6, 57
iris dataset, 126-130

J
JavaScript, warning about, 99
JSON-based HTML, interactivity with, 57
Jupyter Notebooks, Python, 49
    online tutorial with, 57
    origins of, 6, 57
    R similarities to, 16, 22, 110
    VS Code extension for, 16
    for writing reports, 110
Jupyter Project, 6
JupyterLab IDE, 6, 46, 57

K
Keras, Python, 7, 94, 104
key-value pairs, Python, 63
keywords, Python, 59-60, 151

L
L suffix (with integers), R, 25
LabelEncoder, Python, 142
labels
    for plots, 98
    R factor attribute, 32
lapply function, R, 39
"Layered Grammar of Graphics, A" (Wilkinson), 96
layering method of plotting, 96, 137
leaflet package, R, 99-100, 140
levels (groups), 31
LHS and RHS
    in cross-talk script, Python, 118
    lhs:rhs, R, 34
    with operators, 118, 148
.libPaths function, R, 19, 50
libraries and packages, use of terms for, 20, 55
library function, R, 20, 21
line plots for time series data, 87
linear models
    with linear_model, Python, 100
    in Python, 67, 100

About the Authors

Rick J. Scavetta has worked as an independent workshop trainer, freelance data scientist, and cofounder since 2012. Operating as Scavetta Academy, Rick has a close and recurring presence at primary research institutes across Germany. His online courses at DataCamp have been taken by over 200,000 students since 2016, and he's also contributed to advanced data science courses for O'Reilly and Manning. Rick currently serves as the technical curriculum advisor to the Misk Academy in Saudi Arabia and heads the development of their data science program.

Boyan Angelov is a data strategist and consultant with a decade of experience in a variety of academic and industry environments, covering topics such as bioinformatics, clinical trials, HRTech, and management consulting. He is additionally a contributor to open source scientific projects in the field of XAI, and speaks regularly at conferences and meetups.

Colophon

The animal on the cover of *Python and R for the Modern Data Scientist* is a squid (*Loligo forbesii*), a cephalopod commonly found throughout the Atlantic Ocean and along the sea coasts of Western Europe and East Africa. In the winter, squids seek out the stable temperatures on the ocean floor. In the summer, they range anywhere from 33 to 1,640 feet deep.

Squids have eight tentacles below their large eyes and two retractile arms that stretch overhead with suckers on the ends. Above the eyes, two large fins fan out like a diamond. The reddish fins lose color when they jolt backward, using the gills behind the head for jet-propulsion, which helps the squid evade predators. Squids prey upon small fish, crustaceans, and other cephalopods, including fellow squids. They can grow about 14 inches long in their lifespan of 1–2 years.

Many of the animals on O'Reilly covers are endangered; all of them are important to the world.

The cover illustration is by Susan Thompson, based on a black and white engraving from iStock. The cover fonts are Gilroy Semibold and Guardian Sans. The text font is Adobe Minion Pro; the heading font is Adobe Myriad Condensed; and the code font is Dalton Maag's Ubuntu Mono.

O'REILLY®

There's much more where this came from.

Experience books, videos, live online training courses, and more from O'Reilly and our 200+ partners—all in one place.

Learn more at oreilly.com/online-learning

©2019 O'Reilly Media, Inc. O'Reilly is a registered trademark of O'Reilly Media, Inc. 1175

www.ingramcontent.com/pod-product-compliance
Ingram Content Group UK Ltd.
Pitfield, Milton Keynes, MK11 3LW, UK
UKHW011024110225
454912UK00009B/263